TV Antennas and Signal-Distribution Systems

by

M. J. Salvati

Howard W. Sams & Co., Inc.
A Subsidiary of Macmillan, Inc.
4300 West 62nd Street, Indianapolis, Indiana 46268 U.S.A.

FIRST EDITION
THIRD PRINTING — 1985

International Standard Book Number: 0-672-21584-5
Library of Congress Catalog Card Number: 78-64985

Printed in the United States of America.

Preface

A high-quality tv set can reproduce a picture only as good as the signal supplied to it. This means a high-performance tv antenna and signal-distribution system are necessary for full enjoyment of broadcast television. This book describes in detail the requirements, design, and implementation of tv antenna systems capable of providing high-quality tv reception in home and small matv installations.

This book is unique in several aspects. It is the only book below engineering level that provides in-depth information on how tv antennas operate and the performance characteristic of each type of modern tv antenna. It is the only book of any type that also gives performance data on specific antenna models. It is the only book providing performance data based on actual measurements made by the author, rather than simply repeating manufacturers' claims that are often unrealistic. This book was written with the intention that the home owner, tv technician, antenna installer, and motel owner should have the information available that will enable them to make the correct decisions when installing a tv antenna and signal-distribution system.

The subject matter progresses logically all the way from background information on tv reception and antenna basics for the beginner, to how to physically construct the system in the last chapter. In between are chapters full of performance data on over 40 commercial tv antennas and 70 different signal-processing and distribution components, special techniques for peak performance and problem situations, and step-by-step selection and design procedures. All of this information, even the section on antenna principles, is written in plain language. None of the mumbo-jumbo or abstract ideas that normally appear in antenna books will be found here, yet the book is a data gold mine even for experienced antenna men!

M. J. SALVATI

Acknowledgments

This book would not have been possible without the cooperation of a great many people in the industry. I want to thank the following people for their help: Bob Ringland and Red Parsons of Antennacraft; Harry Rheinschmidt of Antenna Corp. of America; Harry Lindblad and Andy Saracena of Astoria Photo; Brad Hubinek and John Young of Belden; Bob Dettman of Blonder-Tongue Labs; Russell Roy and Ira Dryer of Channel Master; Morris Finney Jr. and Richard Phillips of Finney; Ron Rolleri of Gavin; John Bullock and Sudesh Chawla of Jerrold; Jim Sarayiotes and Bob Snow of JFD; Jerry Updyke of Lance Industries; Anne Cominskie of RCA; Arthur Fink, Al Goldstein, and Kerwin McMahon of RMS Electronics; Randy Hibshman of Saxton; and Joe Ridge, John Osborne, and Hal Sorensen of Winegard.

Special thanks to Norman Rea, Chief Antenna Engineer of TACO/Jerrold, and Axel Tyle, Chief Engineer of SITCO, for valuable lessons in the finer points of antenna design, and to my hard-working secretary, Jeannette Pitonzo.

This book is dedicated to
the memory of
Michael Salvati

Contents

CHAPTER 1

ANTENNA SYSTEM BASICS 7

Requirements of a TV Antenna System—TV Bands and Channel Assignments — Signal Level Systems — Decibel Notation — Picture Quality—Reception Range—TV Antenna Principles

CHAPTER 2

VHF ANTENNAS 29

Single-Channel Antennas—Single-Band Antennas—Multiband Antennas—Commercial Multiband Antennas

CHAPTER 3

UHF ANTENNAS 61

Bowtie and Reflector Antennas — Corner-Reflector Antenna — Corner-Reflector Yagi—Log-Periodic Antenna—Hoverman Antenna —Parabolic-Reflector Antenna—Antenna Performance Comparison

CHAPTER 4

ALL-CHANNEL ANTENNAS 85

Isolation Networks—Commercial Antennas

CHAPTER 5

TRANSMISSION LINES 105

Balanced Line—Coaxial Cable—Comparison of Transmission Lines

CHAPTER 6

SIGNAL-DISTRIBUTION COMPONENTS 117

Baluns—Signal Splitters—Band Separators—Directional Couplers—Amplifiers—Attenuators—Miscellaneous

CHAPTER 7

ANTENNA HARDWARE AND ACCESSORIES 151

Masts and Towers—Mounting Brackets and Devices—Guying Hardware—Standoffs—Lightning-Protection Devices—Antenna Rotators

CHAPTER 8

SIGNAL OPTIMIZATION TECHNIQUES 169

Increased Signal Output—Ghost and Interference Elimination—MATV Head Ends

CHAPTER 9

DESIGNING THE ANTENNA SYSTEM 203

Signal Survey—Head-End Design—Distribution-System Design—Matching the Distribution System to the Head End

CHAPTER 10

BUILDING THE ANTENNA SYSTEM 227

Mast and Tower Installation—Installing the Transmission Line—Distributing the Signal

APPENDIX A

ABBREVIATIONS USED IN THIS BOOK 245

APPENDIX B

DIRECTORY OF MANUFACTURERS 247

INDEX 249

CHAPTER 1

Antenna System Basics

This chapter is a "prep course" for the rest of the book. Its purpose is to provide those unfamiliar with antennas with the background information needed to derive the full benefits of the following chapters. As such it outlines the basic requirements of an antenna system, and explains some of the concepts that are elementary to tv antenna systems.

REQUIREMENTS OF A TV ANTENNA SYSTEM

Television systems, whether large or small, consist of the following components:

1. An antenna or antennas that are designed to operate well on the channels available at the reception site. The antenna(s) must deliver sufficient signal output to provide satisfactory picture quality, pick up signals only in the direction of the desired station(s), and reject signals from all other directions.
2. A device for mounting the antenna(s) well above nearby structures.
3. A transmission line that can convey the signal from the antenna to the tv set without excessive signal loss and without any signal pickup of its own.
4. A signal-distribution system that provides the proper amount and quality of signal to each tv set served by it.

TV BANDS AND CHANNEL ASSIGNMENTS

The broadcast stations in the United States operate on frequencies that lie within three bands. These bands are the *vhf* (*very high frequency*) *low band, the vhf high band,* and the *uhf* (*ultrahigh frequency*) *band.* The vhf low band, also referred to as vhf-L in this book, covers the frequency range of 54 to 88 MHz, and contains Channels 2–6. The vhf high band (vhf-H) covers the frequency range of 174 to 216 MHz, and contains Channels 7–13. The uhf band covers the frequency range of 470 to 890 MHz, and contains Channels 14 to 83. However, at present only Channels 14 to 69 (470–806 MHz) are used for prime broadcast; Channels 70 to 83 (806–890 MHz) are reserved for translator use, i.e., the retransmission of other channels to areas which cannot receive them directly.

Table 1-1. Television Channel Assignments

Channel No.	Frequency Limits of Channel (MHz) Lower	Upper
VHF Low Band		
2	54	60
3	60	66
4	66	72
5	76	83
6	82	88
FM Band		
	88	108
Mid Band		
A	120	126
B	126	132
C	132	138
D	138	144
E	144	150
F	150	156
G	156	162
H	162	685
I	168	174
VHF High Band		
7	174	180
8	180	186
9	186	192
10	192	198
11	198	204
12	204	210
13	210	216

Channel No.	Frequency Limits of Channel (MHz) Lower	Upper
Super Band		
J	216	222
K	222	228
L	228	234
M	234	240
N	240	246
O	246	252
P	252	258
Q	258	264
R	264	270
S	270	276
UHF Band		
14	470	476
15	476	482
16	482	488
17	488	494
18	494	500
19	500	506
20	506	512
21	512	518
22	518	524
23	524	530
24	530	536
25	536	542
26	542	548
27	548	554
28	554	560
29	560	566

Table 1-1 cont. Television Channel Assignments

Channel No.	Frequency Limits of Channel (MHz) Lower	Upper	Channel No.	Frequency Limits of Channel (MHz) Lower	Upper
UHF Band					
30	566	572	52	698	704
31	572	578	53	704	710
32	578	584	54	710	716
33	584	590	55	716	722
34	590	596	56	722	728
35	596	602	57	728	734
36	602	608	58	734	740
37	608	614	59	740	746
38	614	620	60	746	752
39	620	626	61	752	758
40	626	632	62	758	764
41	632	638	63	764	770
42	638	644	64	770	776
43	644	650	65	776	782
44	650	656	66	782	788
45	656	662	67	788	794
46	662	668	68	794	800
47	668	674	69	800	806
48	674	680			
49	680	686	Channels 70 through 83 are translator bands, 6 MHz per channel.		
50	686	692			
51	692	698			

Table 1-1 shows the operating frequencies of each individual broadcast channel. You will also note that it contains frequency allocations for other groups of channels in addition to the broadcast channels. These other channels, alphabetically designated, are for nonbroadcast use by catv and matv (community antenna television and master antenna television) systems. The *mid band* covers the frequency range of 120 to 174 MHz, and contains Channels A–I. The *super band* covers the frequency range of 216 to 276 MHz, and contains Channels J–S. These extra channels can be used in several ways. In nearly all catv and in some matv systems, uhf signals picked up by the antennas are converted at the head end to frequencies in the vhf range to minimize signal loss in the cable. Similarly, whenever the catv system provides stations that are broadcast on adjacent channels (e.g., 2 and 3, 10 and 11), one of them is converted to a different channel to minimize adjacent-channel interference on the subscriber's tv set. Also, most catv systems originate programs of their own, continuous weather and time broadcasts, shopping guides, etc. The alphabetical channels are needed when space cannot be found in the regular tv bands, as is usually the case in metropolitan markets.

SIGNAL LEVEL SYSTEMS

The strength or level of a broadcast tv signal at your tv antenna varies according to distance from the transmitter and the intervening terrain. Similarly, the level of the signal delivered to the tv set varies according to the type of antenna, the amount of signal lost in the transmission line, the type of signal-distribution components, etc. Various methods of designating signal level are in use throughout the world. These methods can be classified into two types: voltage-level systems and decibel-level systems.

Voltage Systems

The *volt* (V) is the standard unit of signal level. But the signal levels in tv antenna systems are usually far too small to make the volt a practical unit, so the millivolt (mV) and the microvolt (μV) are most commonly used. The millivolt is one thousandth (0.001) of a volt, and the microvolt is one millionth (0.000001) of a volt, so 1,000,000 μV = 1000 mV = 1 V.

The signal level at the antenna terminals may be as low as a few microvolts in extreme fringe areas to over a hundred millivolts in very-strong-signal areas. Therefore, voltage levels are most often given in microvolts when referring to antenna signal levels, although millivolts are used on occasion. The signal level at any place in a signal distribution system is normally between 1 and 100 millivolts, so millivolts are nearly always used to specify voltage levels in this application.

Decibel System

There is a signal level system that specifies absolute signal level in a way that makes it extremely easy to calculate the effects of component loss on the signal level passing through it. This simplifies system design, as you will see in Chapter 9. This system is the "decibel referred to one millivolt," or *dBmV,* system. This system uses decibel notation to refer all absolute signal levels to one millivolt. For example, 0 dBmV = 1 mV. Table 1-2 shows voltage levels and their equivalents in dBmV. A quick look at this table shows another advantage of dBmV designations; any voltage level from 10 microvolts to 10 volts can be specified approximately using a maximum of three characters (two digits and a polarity sign).

The real advantage of working in decibels rather than in microvolts is that component loss or gain expressed in decibels (dB) is added to or subtracted from the signal levels. This is a lot easier to do (at least in the days before calculators) than the multiplication and division needed where signal levels, gain and loss, etc., are expressed

Table 1-2. Conversion of dBmV and Voltage

dBmV	μV	dBmV	μV	dBmV	mV
−40	10.00	0	1000	41	112
−39	11.22	1	1122	42	125
−38	12.59	2	1259	43	141
−37	14.13	3	1413	44	158
−36	15.85	4	1585	45	177
−35	17.78	5	1778	46	199
−34	19.95	6	1995	47	223
−33	22.39	7	2239	48	251
−32	25.12	8	2512	49	281
−31	28.18	9	2818	50	316
−30	31.62	10	3162	51	354
−29	35.48	11	3548	52	398
−28	39.81	12	3981	53	446
−27	44.67	13	4467	54	501
−26	50.12	14	5012	55	562
−25	56.23	15	5623	56	631
−24	63.10	16	6310	57	707
−23	70.79	17	7079	58	794
−22	79.43	18	7943	59	891
−21	89.13	19	8913	60	1000
−20	100.0	20	10,000	61	1122
−19	112.2	21	11,220	62	1259
−18	125.9	22	12,590	63	1413
−17	141.3	23	14,130	64	1585
−16	158.5	24	15,850	65	1778
−15	177.8	25	17,780	66	1995
−14	199.5	26	19,950	67	2239
−13	223.9	27	22,390	68	2512
−12	251.2	28	25,120	69	2818
−11	281.8	29	28,180	70	3162
−10	316.2	30	31,620	71	3548
− 9	354.8	31	35,480	72	3981
− 8	398.1	32	39,810	73	4467
− 7	446.7	33	44,670	74	5012
− 6	501.2	34	50,120	75	5623
− 5	562.3	35	56,230	76	6310
− 4	631.0	36	63,100	77	7079
− 3	707.9	37	70,790	78	7943
− 2	794.3	38	79,430	79	8913
− 1	891.3	39	89,130	80	10,000
0	1000.0	40	100,000		

in voltage terms. The use of the decibel and the dBmV has therefore become widespread in the tv antenna and matv field.

DECIBEL NOTATION

The decibel, on which the dBmV system is based, is a convenient way of referring to the *ratio* between two quantities, such as the input and output levels of an amplifier (amplifier gain), the output of one antenna relative to another (antenna gain), the magnitude of the signal pickup at the rear of an antenna compared to the signal pickup at its front (front-to-back ratio), or the ratio of the noise in a signal to the amount of the desired signal. To better understand the significance of the dBmV voltage level system described in the preceding section, and the use of decibels in the following sections, let us study decibels a bit more.

The decibel is a "shorthand" method of describing the ratio between two power levels, or two voltage levels in the same impedance (this term will be defined later). Table 1-3 shows the decibel equivalents of different voltage and power ratios. Notice that 3 dB represents a power ratio of 2:1, and a voltage ratio of 1.4:1. There is no contradiction here because a voltage increase of 41 percent across a certain impedance results in twice the power developed in that impedance. This particular ratio, 3 dB, was selected for this example because it is very significant; each time you double the number of antennas in an array, gain increases by 3 dB.

Decibel notation is very handy for calculating total loss in a complex system and the signal levels that result. For instance, if we start out with 2000 microvolts (+6 dBmV) at the antenna, run the signal through a transmission line that has a 2-dB loss, then through a 15-dB amplifier, and then divide this with a power splitter having a 3-dB loss, we can mentally calculate the final signal level as follows:

$$+6 - 2 + 15 - 3 = +16 \text{ dBmV}$$

This is a lot faster than trying to do the following:

$$2000 \times 0.794 \times 5.62 \times 0.708 = ???$$

PICTURE QUALITY

The picture on a high-quality, properly adjusted tv receiver can look almost as good as a well-made photograph when viewed from the proper distance, and with the tv receiver fed a perfect signal. Unfortunately, perfect signals are hard to obtain. Broadcast signals may suffer from one or two problems: insufficient signal level or ghosts. Either one of these problems reduces picture quality; if severe enough either one can make the picture unusable.

Ghosts

The signal radiated by a tv transmitting antenna may arrive at the receiving antenna by more than a single path. This is because tv signals are reflected by large metallic objects (such as water tanks and steel-frame buildings) and even by many natural objects (such

Table 1-3. Decibel Conversion

Voltage Ratio*	Power Ratio	dB	Voltage Ratio*	Power Ratio
1.000	1.000	0	1.000	1.000
0.989	0.977	0.1	1.012	1.023
0.977	0.955	0.2	1.023	1.047
0.966	0.933	0.3	1.035	1.072
0.955	0.912	0.4	1.047	1.096
0.944	0.891	0.5	1.059	1.122
0.933	0.871	0.6	1.072	1.148
0.923	0.851	0.7	1.084	1.175
0.912	0.832	0.8	1.096	1.202
0.902	0.813	0.9	1.109	1.230
0.891	0.794	1.0	1.122	1.259
0.841	0.708	1.5	1.189	1.413
0.794	0.631	2.0	1.259	1.585
0.750	0.562	2.5	1.334	1.778
0.708	0.501	3.0	1.413	1.995
0.668	0.447	3.5	1.496	2.239
0.631	0.398	4.0	1.585	2.512
0.596	0.355	4.5	1.679	2.818
0.562	0.316	5.0	1.778	3.162
0.531	0.282	5.5	1.884	3.548
0.501	0.251	6.0	1.995	3.981
0.473	0.224	6.5	2.113	4.467
0.447	0.200	7.0	2.239	5.012
0.422	0.178	7.5	2.371	5.623
0.398	0.159	8.0	2.512	6.310
0.376	0.141	8.5	2.661	7.079
0.355	0.126	9.0	2.818	7.943
0.335	0.112	9.5	2.985	8.913
0.316	0.100	10	3.162	10.00
0.282	0.0794	11	3.55	12.6
0.251	0.0631	12	3.98	15.9
0.224	0.0501	13	4.47	20.0
0.200	0.0398	14	5.01	25.1
0.178	0.0316	15	5.62	31.6
0.159	0.0251	16	6.31	39.8
0.141	0.0200	17	7.08	50.1
0.126	0.0159	18	7.94	63.1
0.112	0.0126	19	8.91	79.4
0.100	0.0100	20	10.00	100.0
3.16×10^{-2}	10^{-3}	30	3.16×10	10^{3}
10^{-2}	10^{-4}	40	10^{2}	10^{4}
3.16×10^{-3}	10^{-5}	50	3.16×10^{2}	10^{5}
10^{-3}	10^{-6}	60	10^{3}	10^{6}
3.16×10^{-4}	10^{-7}	70	3.16×10^{3}	10^{7}
10^{-4}	10^{-8}	80	10^{4}	10^{8}
3.16×10^{-5}	10^{-9}	90	3.16×10^{4}	10^{9}
10^{-5}	10^{-10}	100	10^{5}	10^{10}
3.16×10^{-6}	10^{-11}	110	3.16×10^{5}	10^{11}
10^{-6}	10^{-12}	120	10^{6}	10^{12}

* For comparing voltages in equal impedances.

Fig. 1-1. Television picture with multipath interference.

as mountains and certain air masses). This is quite common in cities, where many tall buildings are clustered, but it can also occur in remote mountainous areas and canyons. The reflected signals are properly called multipath signals, although they are most often called *ghosts* because of their effect on the tv picture.

Ghosts cause a problem because the lengths of the multipath signals are always longer than the path of the direct signal (the shortest distance between two points being a straight line). The extra distance means the reflected signals reach the tv set a little after the direct signal, and multiple images result (Fig. 1-1). Also, each time a signal is reflected, a 180° phase reversal occurs; so ghosts may partially cancel the direct signal.

As Fig. 1-2 shows, multipath signals may hit the receiving antenna from any direction relative to the transmitter—front, side, or

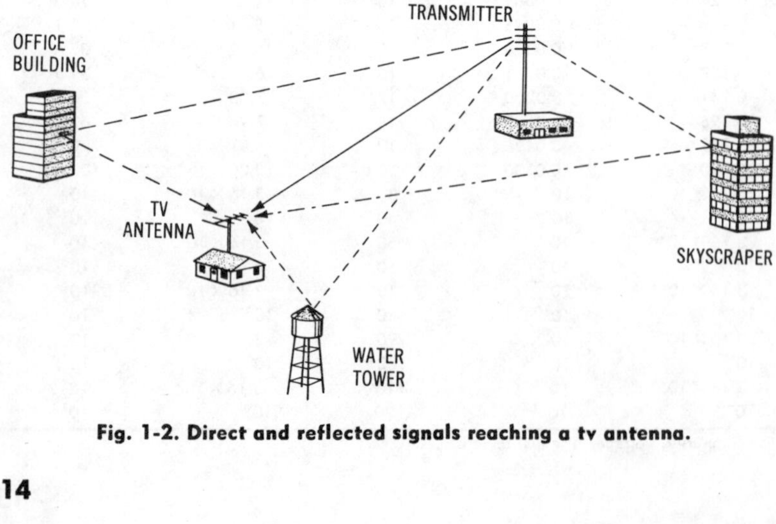

Fig. 1-2. Direct and reflected signals reaching a tv antenna.

back. In some circumstances there is no way to eliminate ghosting. In most cases, however, ghosting can be either completely eliminated or greatly reduced by using the right antenna and/or the techniques described in Chapter 8.

Signal-to-Noise Ratio

If you tune in to a channel where there is no signal broadcast, you will not see a blank screen and hear no sound. Sparkling and ever-changing specks commonly called "snow" will appear on the screen, and a rushing noise will be heard. What you are seeing and hearing is the internal noise of the tv set. In the *absence* of a signal, the amplifiers in the tv set operate at maximum gain, and the noise generated by the first transistor in the tuner is amplified by the rest of the circuit to the level that appears on the tv screen. When a *weak* signal is present, the tv operates at somewhat less gain, so internal noise is not amplified as much and is also masked somewhat by the signal. When a *strong* signal is received, the tv operates at such low gain that the internal noise cannot be amplified enough to be noticed at all. The degree to which noise in the tv picture is noticeable is called *signal-to-noise-ratio,* and is abbreviated *snr* or *S/N ratio.*

The relationship between perceived picture quality and noise level depends on the judgment of the viewer as to what constitutes a good picture. By means of a survey of a great number of "average" viewers watching a variety of tv sets, the relationships shown in Table 1-4 were determined. An *excellent* or *Grade A* picture has absolutely no perceptible noise. A *fine* or *Grade B* picture has noise that is perceptible if looked for, but is not normally noticed. A *passable* or *Grade C* picture (Fig. 1-3) has a very noticeable amount of noise,

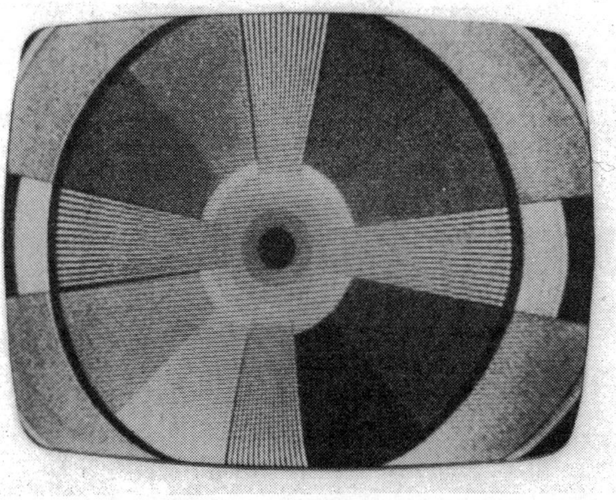

Fig. 1-3. Grade C picture.

but not so much as to interfere with viewing the program. An example of a picture too snowy to be enjoyable is shown in Fig. 1-4.

The signal levels required at the *75-ohm* input of the typical tv set to produce these signal-to-noise ratios are also given in Table 1-4. *Double these levels* are required at *300-ohm* input terminals. Twice as much signal is required at the uhf antenna terminals to produce the same picture quality as on vhf because the uhf tuner used in the average tv set generates a lot more internal noise than does the vhf tuner.

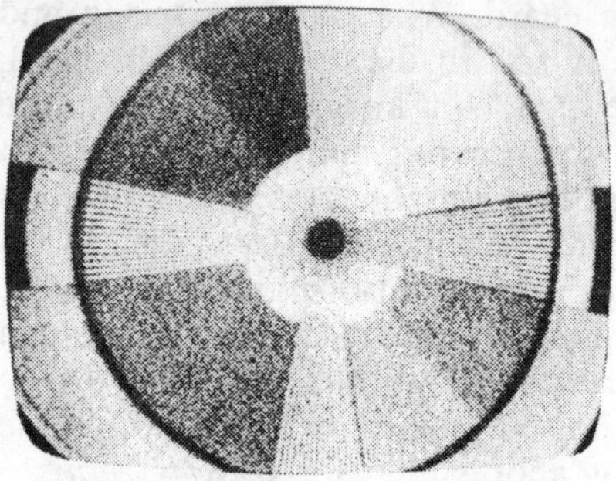

Fig. 1-4. Poor (very snowy) picture.

Table 1-4. Signal Level vs. Picture Quality

Picture Quality	SNR in dB at TV Set	Required Signal Level* in μV VHF	UHF
Excellent/Grade A	44	500	1000
Fine/Grade B	34	150	300
Passable/Grade C	28	90	180
Poor	20	40	80

* For 75-ohm input.

RECEPTION RANGE

The question most often asked by antenna purchasers is something to the effect of "what's the range of this antenna?" Unfortunately there is really no accurate answer to this question. Reception range depends on the following factors: transmitter power, transmitting-antenna gain, transmitting-antenna height, receiving-antenna gain, receiving-antenna height, and the intervening terrain and structures. The only one of these factors that antenna manufacturers have any control over is receiving-antenna gain, yet a good many manufacturers list distance figures in their advertising. Some qualify the advertisement with the words "over favorable terrain," others do not. In nearly all cases the figures represent more wishful thinking than anything else, with claims of 100 miles for medium-sized vhf antennas and 175 miles for the very large ones. Reception is indeed possible at these distances, but Grade A pictures are highly unlikely. A more reasonable estimate is a Grade B or C picture, depending on the terrain. For this reason the Grade A range estimates assigned by the author to the various antenna models in Chapters 2–4 are often but a fraction of the manufacturers' estimates. These figures are for convenience in *comparison only,* and are not to be taken as guarantees of perfect reception, or as limits on the effective ranges of the antennas.

The field strength of broadcast tv signals falls off roughly as the square of the distance from the transmitter. This means that if a particular antenna model had an output of 8000 microvolts at a distance of 10 miles from the transmitter, the same antenna model located 20 miles from the transmitter would have an output of about 2000 microvolts. At 40 miles the output level would be only 500 microvolts, at 80 miles only 125 microvolts. This assumes all other factors (such as antenna height above surrounding terrain and average height above sea level) are equal and that there are no large obstructions on the intervening terrain (smooth earth). In real life things are never this uniform, so levels at each point in this example would probably differ somewhat. The point is that field strength, hence antenna output, falls off rapidly with distance, and those living over 70–80 miles from the transmitter (or at shorter distances with poor terrain) will find it difficult to achieve the signal levels needed for a Grade A picture.

TV ANTENNA PRINCIPLES

To derive the maximum benefit from the antenna descriptions in the following chapters, it is necessary to have a rough understanding of a few antenna concepts, or at least the terms used to describe them. The explanations of these terms use simplifications of true antenna theory to present some rather abstract concepts in a form most useful to the readers of this book. Moreover, just enough theory is included to serve the purpose of the book; this chapter is not intended as a general course in antenna theory.

Frequency and Wavelength

Because of the antenna condition known as resonance (described next), antenna dimensions are related to wavelength. Theoretically, a wavelength is the distance covered by electromagnetic radiation in the time it takes for it to undergo one complete cycle (two changes of polarity).

Wavelength is measured in some convenient distance measure: feet, inches, meters, etc. Wavelength can be calculated from the frequency, and vice versa, by any of the following formulas. (The

Greek letter lambda, λ, is used as the symbol of wavelength in formulas.)

$$\lambda_{feet} = \frac{984}{f_{MHz}} \qquad \lambda_{inches} = \frac{11{,}808}{f_{MHz}} \qquad \lambda_{meters} = \frac{300}{f_{MHz}}$$

For example, 57 MHz (middle of Channel 2) has a wavelength of 17.3 feet, while 213 MHz (middle of Channel 13) has a wavelength of only 4.6 feet. Notice that as the frequency increases, wavelength decreases.

Resonance

Most antenna dimensions are critical because antennas are frequency-sensitive devices. This means than an antenna designed for one frequency or band will have different dimensions than an antenna designed for a different frequency or band, though the antenna configurations may be the same.

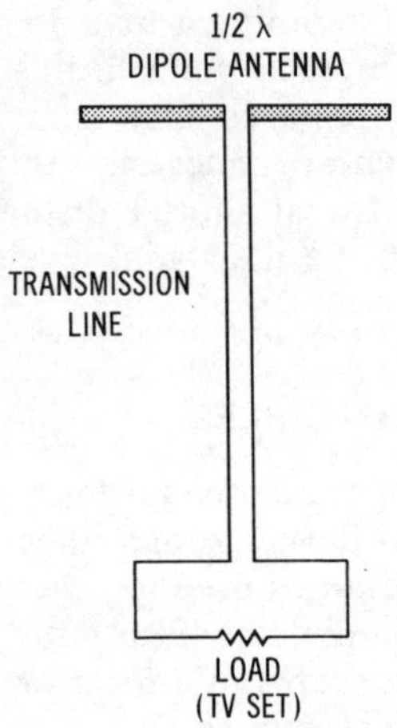

Fig. 1-5. Dipole antenna.

The reason antenna dimensions are related to frequency and wavelength is the condition known as *resonance*. Resonance occurs when the length of an antenna element is an exact multiple of an electrical half wavelength (½ λ) at the frequency being received. (You might say the radio wave fits the antenna element just right!) Since the purpose of an antenna is to deliver the maximum-possible power from the intercepted radio (tv) wave for the antenna length and field strength involved, resonance allows the radio waves to induce the maximum current in the element. The shortest element length that can resonate at a particular frequency is one-half wavelength long at the frequency. This length and one wavelength are the element lengths most often used for tv antennas.

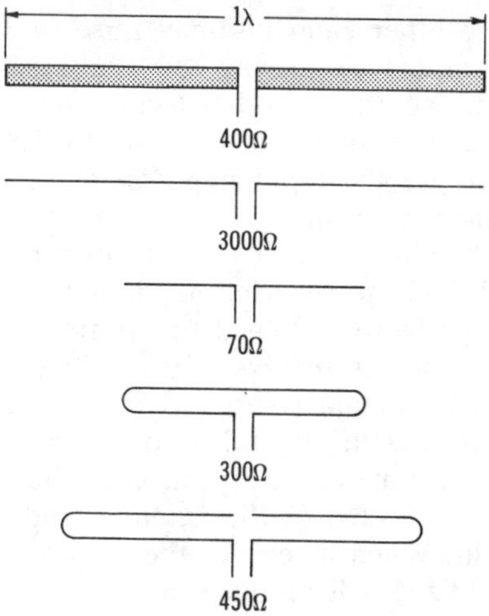

Fig. 1-6. Factors affecting dipole impedance.

If a resonant element is split in the middle and a transmission line attached (Fig. 1-5), the antenna element (called a *dipole*) can deliver a significant amount of signal power to a suitable load at the other end of the transmission line. A tv set or the input of a signal-distribution system constitutes a suitable load. *Maximum*-possible power transfer to a load occurs when that load is the same *impedance* as the dipole.

Impedance

As stated before, the purpose of an antenna is to deliver signal power to the tv receiver. The terminal impedance of an antenna or dipole element can be described simply as the relationship of current and voltage in the power output of the antenna or dipole.

The terminal impedance of a dipole element at its resonant frequency depends on three factors. They are its length (in terms of wavelength), its length-to-diameter ratio, and its configuration. As Fig. 1-6 shows, a thin one-wavelength dipole has an impedance very much higher than that of a half-wavelength dipole. Fat elements have lower impedance than skinny elements of the same length and configuration. Although the impedance of an infinitely thin (theoretical) one-wavelength dipole is too high to be practical, the thick tubing from which tv antenna elements are constructed results in a one-wavelength dipole with a 400- to 500-ohm impedance. A *folded* dipole has four times higher impedance (300 ohms) than the standard dipole, and a folded dipole three-quarters wavelength long has still higher terminal impedance (450 ohms).

The aforementioned impedances are for a solitary dipole. Television antennas, however, have many elements. The proximity of these other elements may raise or lower the terminal impedance of the dipole(s) driving the transmission line. Antenna designers use these factors and other techniques to produce an antenna whose terminal impedance is as close to 300 or 75 ohms as possible, since these are the input impedances of tv sets and amplifiers, and of tv transmission lines.

Matching antenna, transmission line, and load serves a purpose additional to securing maximum signal-power transfer. If the load is not perfectly matched to the transmission line, a portion of the signal power is *reflected* by the load and travels back to the antenna. If the antenna *is* properly matched to the line, no problem occurs other than a slight loss of power. However, if the antenna is not properly matched, a portion of the signal reflected from the load will be re-reflected by the antenna, and travel down the transmission line again. When it reaches the load a second time, the portion of it that is utilized will produce the same picture as the first-time signal. This appears as a ghost in the picture. The separation between the two pictures is dependent on the time it takes to travel up the transmission line and down again, hence it is proportional to line length. With long transmission lines the separation is great enough to be noticeable.

Directionality

"Directionality" refers to the ability of an antenna or antenna element to pick up signals in one direction relative to another. It is generally desirable for an antenna to be able to pick up signals extremely well in one direction and very poorly in all other directions. This allows the desired station to come in well, and many forms of interference (including other tv stations) to be rejected. It improves the signal-to-noise ratio of the picture and minimizes ghost pickup as described earlier.

Antenna directionality is best shown by means of *polar patterns.* These patterns are plots of the relative *voltage response* or signal pickup at all directions relative to the desired direction. The overall (three-dimensional) antenna pattern is customarily broken down into a pair of two-dimensional plots: the *horizontal-plane* polar pattern and the *vertical-plane* polar pattern. The horizontal-plane pattern shows N-E-S-W directions. This is the most important of the patterns, hence is the one used in this book to show antenna directionality. The vertical-plane polar pattern shows up and down response. Since response in this plane is greatly affected by antenna height in practical installations, and is important only in certain interference situations, the vertical-plane polar pattern is rarely shown.

An antenna's horizontal-plane polar pattern is greatly affected by the directionality of the individual elements that make up the an-

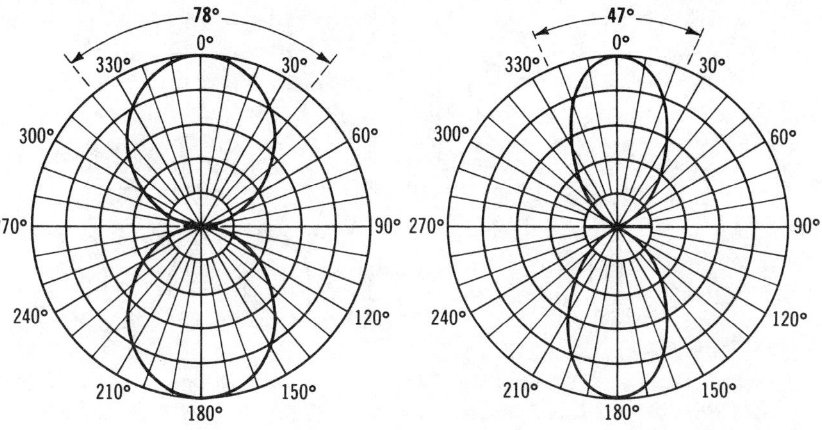

(*A*) *For ½-wavelength dipole.* (*B*) *For 1-wavelength dipole.*

Fig. 1-7. Polar patterns of dipoles.

tenna. A half-wavelength element has a "figure-eight" polar pattern, like that shown in Fig. 1-7A. Each bulge in the pattern is called a *lobe.* A one-wavelength element also has a two-lobe pattern (Fig. 1-7B), but the lobes are much narrower. As you will find out later, this indicates potentially higher antenna gain. But for now the important thing to consider is the reduced pickup in directions other than the front-to-back direction. Notice that at 30° 150°, 210°, and 330° the response is only 57 percent for the full-wave dipole compared with 85 percent for the half-wave dipole. This *front-to-side ratio* (fsr) is conveniently expressed in decibels. Since the response of the full-wave dipole at 30° is 4.9 dB below maximum (see Table 1-3), and the response of the half-wave element is 1.4 dB below maximum, the full-wave dipole has 3.5 dB better front-to-side ratio than the half-wave dipole at the 30° bearing.

At the next resonant length (1½ wavelengths) the single front and back lobes that characterized the half- and one-wavelength dipoles have split into three lobes each (Fig. 1-8A). Pickup is no longer maximum in a direction perpendicular to the element; the side lobes are larger than the front and back lobes. At higher multiples of a half-wavelength (Fig. 1-8B for example), the number and size of the side lobes increase, and pickup is maximum in a direction approaching that of the element. Because of this undesired pickup from the sides, and poor pickup from the front, antenna elements longer than one wavelength are not used without corrective techniques (discussed in Chapter 2).

The points at which the antenna voltage response is 3 dB below (0.7 times) maximum is a special case. These are the *half-power*

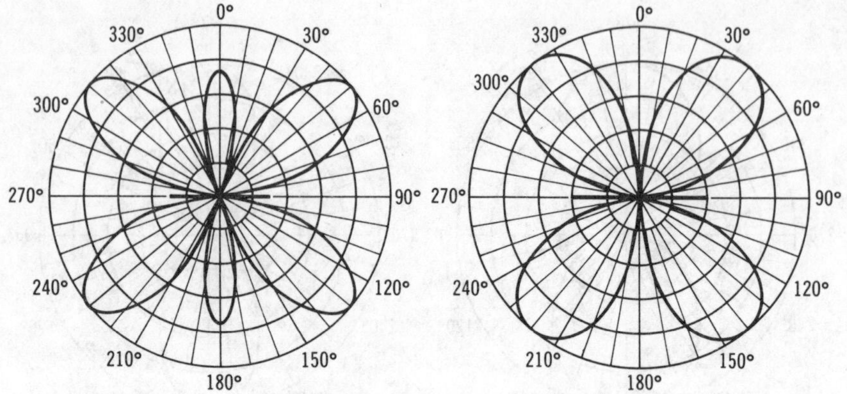

(*A*) *For 1½-wavelength dipole.* (*B*) *For 2-wavelength dipole.*

Fig. 1-8. Polar patterns of 1½- and 2-wavelength dipoles with large sidelobes.

points, and the angular distance (degrees) between the two bearings where this occurs on the main lobe is called the *beamwidth* of the antenna. For example, the −3-dB beamwidth of the half-wave dipole in Fig. 1-7A is 78°, and that of the full-wave dipole in Fig. 1-7B is 47°.

The polar patterns provided by most manufacturers show the voltage response of the antenna on a purely arbitrary 0–100 scale. The response of the antenna in the forward direction is always set at 100 regardless of the antenna gain, so polar patterns give no gain information on antennas, just directionality.

Bandwidth

Antenna resonance, gain, impedance, pattern, etc., are frequency-sensitive conditions. As you move away in frequency from the resonant frequency of an antenna (where the element length is a half wavelength, one wavelength, etc.), the antenna characteristics change. The degree to which they change depends on the antenna's *bandwidth.* The bandwidth of an antenna may be considered the range of frequencies on either side of the resonant frequency where the antenna characteristics are close enough to those at resonance that the antenna performs pretty much the same.

An antenna with narrow bandwidth performs well only over a relatively small frequency range. This is why single-channel antennas are called *narrow-band* antennas. An antenna with wide bandwidth performs well over a relatively wide frequency range. Since the tv bands cover very broad frequency ranges, general-purpose tv antennas must have very wide bandwidth.

There are many techniques used to produce antennas with wide bandwidth. Some of these techniques are discussed in the chapters on antenna types. Right now we will consider what is possibly the most basic technique, the use of very thick antenna elements. A very thin antenna element, like the dipole shown in Fig. 1-9A, has characteristics that remain substantially constant only over a frequency range that is a small percentage of its resonant frequency. At frequencies above and below this narrow frequency range, the antenna behaves very differently. The antenna element shown in Fig. 1-9B is approximately the same length, and has the same resonant frequency. However, its diameter is now a significant percentage of its length. The characteristics of this antenna element remain substantially the same over a much wider frequency range above and below its resonant frequency. This is a *wideband,* or *broadband,* element. At uhf frequencies much smaller length-to-diameter ratios are possible because the element lengths are much shorter *and*

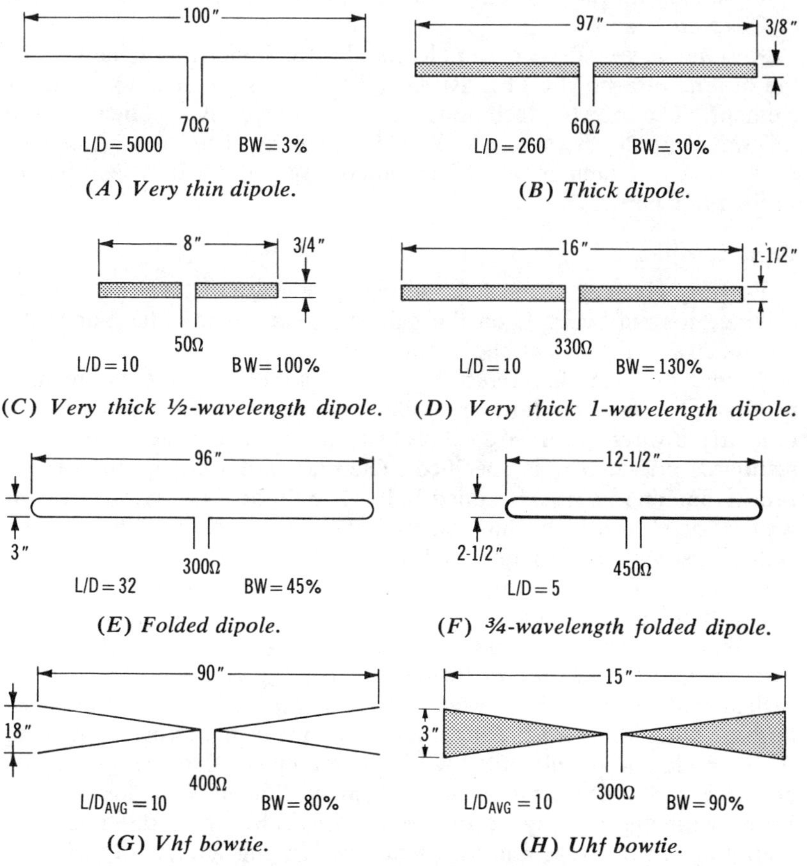

(A) Very thin dipole.

(B) Thick dipole.

(C) Very thick ½-wavelength dipole.

(D) Very thick 1-wavelength dipole.

(E) Folded dipole.

(F) ¾-wavelength folded dipole.

(G) Vhf bowtie.

(H) Uhf bowtie.

Fig. 1-9. Bandwidths of various dipoles.

thicker tubing is practical. Therefore, the uhf dipole shown in Fig. 1-9C has even wider bandwidth. Since 50 ohms is too low for tv antenna use, this technique is most often applied to full-wave uhf dipoles (Fig. 1-9D). The length-to-diameter ratio of 10 lowers the 3000-ohm impedance of the theoretical full-wave dipole down to only 300–400 ohms, and produces an extremely wide bandwidth.

Length-to-diameter ratios of 10 are not practical at vhf frequencies, so the *folded dipole* (Fig. 1-9E) is used in some designs to produce a single dipole with a very wide bandwidth. By making the spacing between the parallel portions of the element wide, the equivalent of a very thick dipole can be built using small-diameter tubing. This technique can be extended to the three-quarters wavelength folded dipole shown in Fig. 1-9F. This dipole is frequently used in uhf antennas because of its favorable terminal impedance and open top. Commercial uhf and vhf-uhf antennas using this dipole are described in Chapters 3 and 4.

Another technique occasionally used is to build an element along the outline of a bowtie (Fig. 1-9G). This also simulates a very thick element. The bowtie technique is very frequently applied to uhf full-wave dipoles, either in outline form or as sheet-metal elements (Fig. 1-9H). Commercial uhf antennas using these bowties are described in Chapter 3.

Gain

For any given tv-signal field strength, the output level of a tv receiving antenna depends on the gain of the antenna. The higher the antenna gain, the higher the output level.

Gain is a term that refers to the output of an antenna relative to that of some standard antenna. The ratio of the output levels is nearly always given in decibels. Any antenna can be used as the standard, provided it is specified. The standard most often used to rate tv antenna gain is a "tuned half-wave dipole" or "resonant half-wave dipole" for each channel, working into a matched load. Gain stated this way is often qualified as "dB ½ λ." Absolute gain statements in this book use this reference. However, some manufacturers make their antenna-gain claims sound better by stating gain relative to a theoretical antenna called an isotropic radiator. Gain stated this way, which should be qualified by "dBi," gives a number 2 dB higher than when a half-wave dipole is the standard. Always determine what standard the manufacturers are using before comparing the gain claims for different brands of antennas. Also, be aware that accurate gain measurements are extremely difficult to make, so published gain figures may be off by ±1 dB. This, plus differences in measurement facilities and technique, adds a big uncertainty to comparing one manufacturer's gain claims to those of another.

Antenna gain is determined primarily by the three-dimensional pickup pattern of the antenna. Higher gain is achieved by concentrating the pickup pattern in one direction. Therefore, antenna gain varies inversely with beamwidth (in both horizontal and vertical planes). This means the narrower the beamwidth, the higher the gain, when all other factors are constant.

Television gain depends on the antenna type, size, and band(s) covered. These factors result in gain figures of anywhere from 0 to 18 dB. Note that having a gain of 0 dB on some channel does not mean that an antenna has no output, but merely that it is the same amount (1.0 times) as the output from a half-wavelength dipole resonant on that channel. A negative decibel figure means the antenna output is *less* than that of a resonant half-wave dipole.

The gain of the typical broadband tv antenna is not constant across the tv band for which it is designed. Unless the gain distribution is peculiar, this is usually not too important. In fact, a gain characteristic that rises with frequency is *usually* desirable. The important factor is that the gain is reasonably even or *flat* across each channel. A top-notch antenna will have no more than 1-dB gain change across each 6-MHz channel width; a fair antenna may have as much as 2-dB gain variation across some channels. Small vhf and vhf-uhf antennas generally have poor gain curves, full of sizable peaks and dips, with variations of 3 dB or more across some channels. This is one of the reasons small antennas are not recommended even if the signal is very strong.

Parasitic Antennas

In all the preceding discussions of antenna characteristics a single element or dipole was used in the explanation. Although a single dipole can be (and occasionally is) used for television reception, high-performance television antennas need much more than one element.

Nearly every vhf, vhf-uhf antenna, and uhf antenna uses *parasitic* elements. A parasitic element is not electrically connected to the transmission line, as is the *driven* element(s). Parasitic elements increase the signal current in the driven element, hence the gain of the antenna, by virtue of their proximity effects. Parasitic elements are always a little shorter or a little longer than a half-wavelength. If the element is shorter than an electrical half-wavelength, it is called a *director*. If it is longer than an electrical half-wavelength, it is called a *reflector*. Directors *increase* the antenna pickup on the side on which they are located; reflectors *decrease* the antenna pickup on their side. A reflector and one or more directors can therefore change the bidirectional pattern of the simple half-wavelength dipole (Fig. 1-7A) to one that is largely unidirectional and thereby increase the antenna

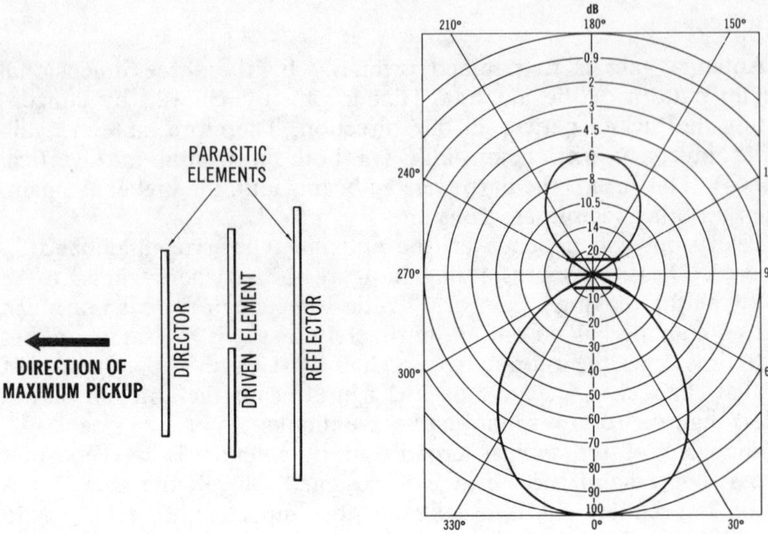

(*A*) *Elements of yagi.* (*B*) *Polar pattern of yagi.*

Fig. 1-10. Simple yagi antenna and its pattern.

gain in that direction. An antenna of this type is called a parasitic, or *yagi,* antenna (Fig. 1-10).

Since the front and rear lobes of the yagi antenna are of different sizes, the *front-to-back ratio* (fbr) is greater than 1. In the example shown in Fig. 1-10B, the relative pickup of the front lobe is 100 units, and that of the back lobe is 50 units. The front-to-back ratio is always expressed in decibels, so the fbr 100/50 is expressed as 6 dB. Unidirectional pickup is generally highly desirable, so a high fbr is an indicator of a good antenna.

Commercial tv antennas using yagi design principles always have more than the two parasitic elements shown in Fig. 1-10A. The usual arrangement in vhf antennas is one reflector and several directors. Broadband uhf antennas also use one reflector and many directors, but the "one reflector" usually consists of a wire grid or number of

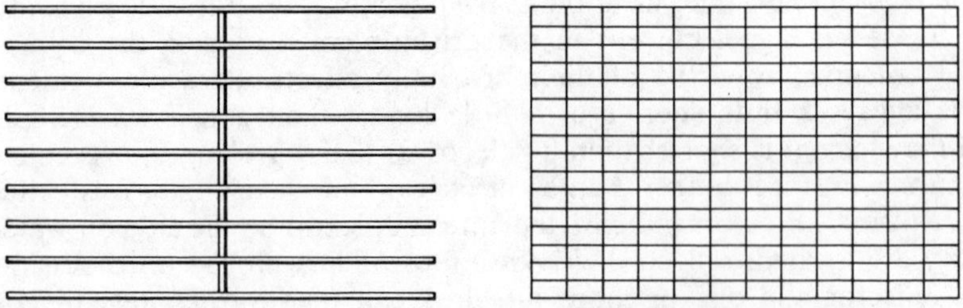

(*A*) *Rod type.* (*B*) *Wire-grid type.*

Fig. 1-11. Broadband plane reflectors.

(A) With vhf-H directors in vertical plane.

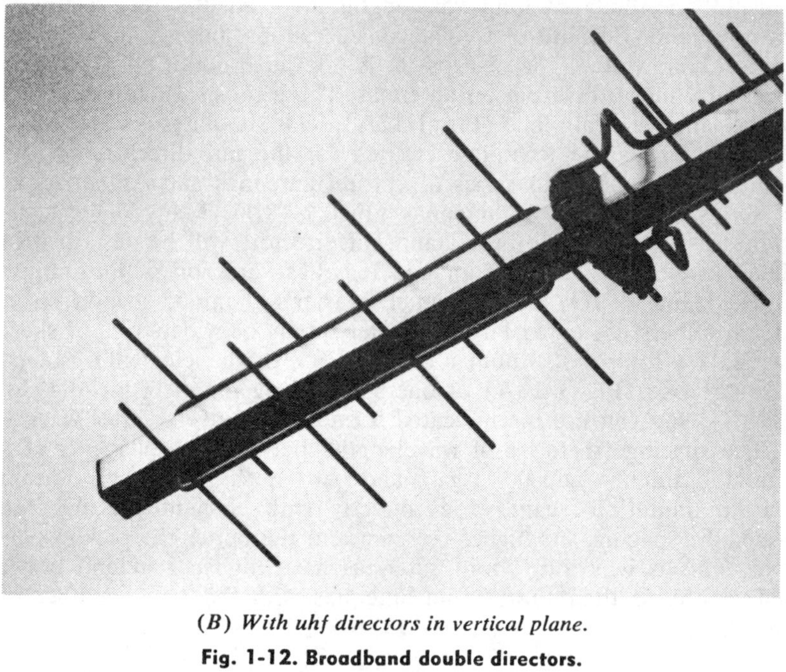

(B) With uhf directors in vertical plane.

Fig. 1-12. Broadband double directors.

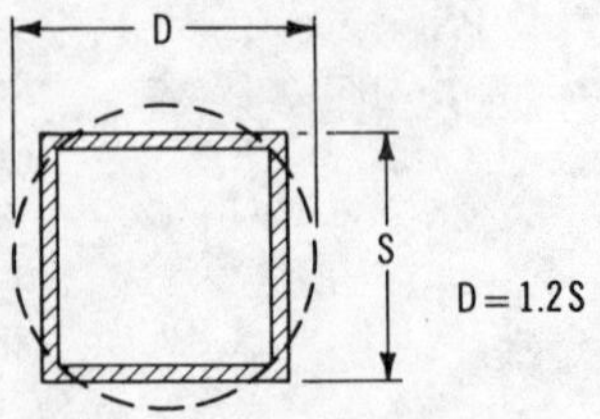

(*A*) *Square tube.*

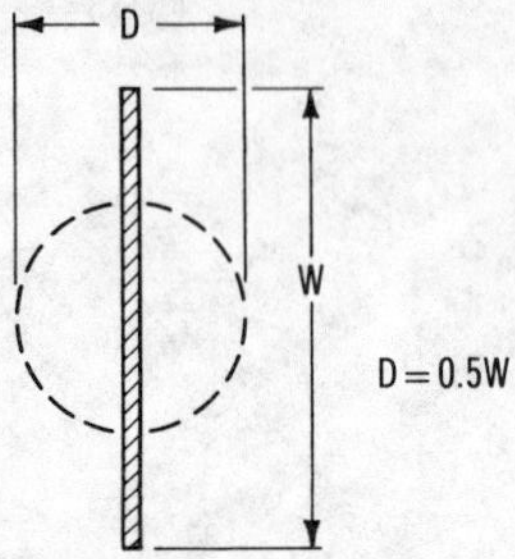

(*B*) *Flat strip.*

Fig. 1-13. Tubular equivalents.

closely spaced rods arranged in a flat plane (Fig. 1-11). A plane reflector that is at least a half wavelength high and one wavelength wide at the lowest frequency covered by the antenna produces much higher gain and front-to-back ratio than a single-rod reflector. Moreover, its effects remain constant over a very wide range of frequencies, unlike the performance of a single-rod reflector. The plane reflector is therefore a broadband element.

A similar situation exists in regard to the directors and bandwidth. Since vhf directors have relatively high length-to-diameter ratios because of the relatively long wavelengths involved, they are relatively narrow band. Because of weight considerations, little is done to increase bandwidth. The exception is the high-band vhf directors used on Winegard Chromstar antennas. These consist of two directors in the same vertical plane (Fig. 1-12A). This simulates a single thick director. The same technique is used for the uhf directors on the Winegard Premier (X) series of vhf-uhf antennas and Antennacraft Y series of log-periodic antennas (Fig. 1-12B). Many other techniques used to produce broadband uhf directors will be described in detail in Chapter 3. These include Jerrold's variation of the outline bowtie (Fig. 1-9G) and Channel Master's diamond-shaped sheet metal. Others use either large-diameter tubing or wide strips of sheet metal. To compare a tubular director (or driven element) against a square one (Fig. 1-13A) or one stamped out of sheet metal (Fig. 1-13B), you can use the indicated formula to convert either way.

The spacing (in terms of wavelength) between the elements of a parasitic antenna affects its gain and pattern characteristics. Within certain limitations, gain varies directly with element spacing: the wider the spacing, the higher the gain and the better the pickup pattern. This is why long-boom antennas have higher gain and better polar patterns than short-boom antennas with the same number of elements.

CHAPTER 2

VHF Antennas

The vhf tv channels are the mainstay of commercial broadcast television, so the selection of the proper vhf antenna is essential for most people for tv enjoyment.

This chapter discusses three categories of vhf antennas: single-channel, single-band, and multiband antennas. Representative models of each type were selected for presentation in this chapter. Of these, the multiband (12-channel) antenna is the most commonly used, hence the most important.

The number of multiband antenna models offered by antenna manufacturers totals well over a hundred. The author has tested quite a few of the small- to medium-sized models; those that worked especially well are featured in the latter part of this chapter.

All of the acceptable multiband antennas and most of the single-band antennas are log-periodic derivatives. The older simpler types, such as the flying V, conical, and simple yagi antennas, did not meet the author's performance standards, so they are not covered in this book. Similarly, the two- and three-element "little brothers" of the high-performance log-periodics are not included either because they do not have sufficiently good polar patterns or gain flatness for a quality antenna system.

SINGLE-CHANNEL ANTENNAS

The antenna type having the greatest gain, highest front-to-back ratio, and the narrowest beamwidth for its weight and cost is the single-channel antenna. Since the bandwidth required for a single-channel antenna is only 6 MHz (compared to 162 MHz for an all-vhf-channel antenna), gain can be very high.

The yagi antenna is invariably the design chosen for single-channel antennas because a large number of directors produce a very high gain for the antenna size over a narrow bandwidth. A folded dipole is often used as the driven element to obtain the needed 6-MHz bandwidth and to maintain a high antenna impedance. This approach is used in the SITCO CA line of high-performance single-channel catv antennas. The eight-element low-band models yield 14 dB gain, about 22 dB front-to-back ratio, and have very narrow beamwidth (Fig. 2-1 top). Since the 6-MHz channel beamwidth is a significant percentage of the antenna center frequency in the vhf low band, the gain flatness over the channel width is around 2 dB on the low-

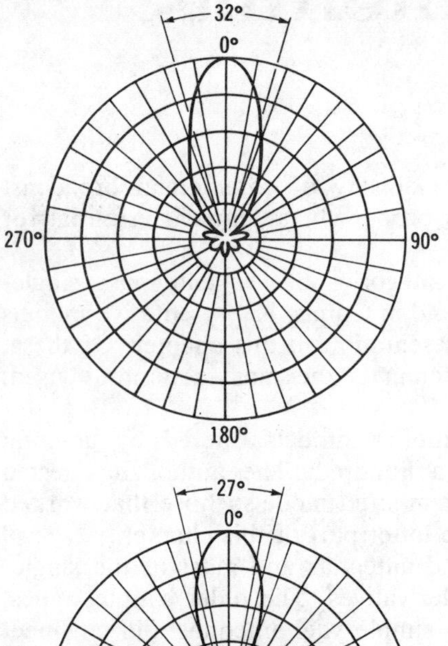

(A) For SITCO CA 8-1 antenna.

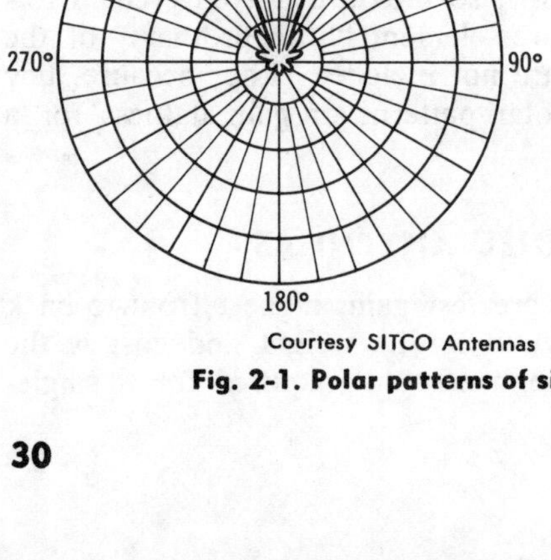

(B) For SITCO CA 12-1 antenna.

Courtesy SITCO Antennas

Fig. 2-1. Polar patterns of single-channel antennas.

channel antennas. The 12-element high-band models (Fig. 2-2) produce 15 dB gain, a phenomenal 28-dB front-to-back ratio, and even narrower beamwidth! Because the 6-MHz channel width is a very small percentage of the antenna center frequency on the high-band channels, the gain flatness of the CA 12-1 series antennas is about 1 dB over the channel width.

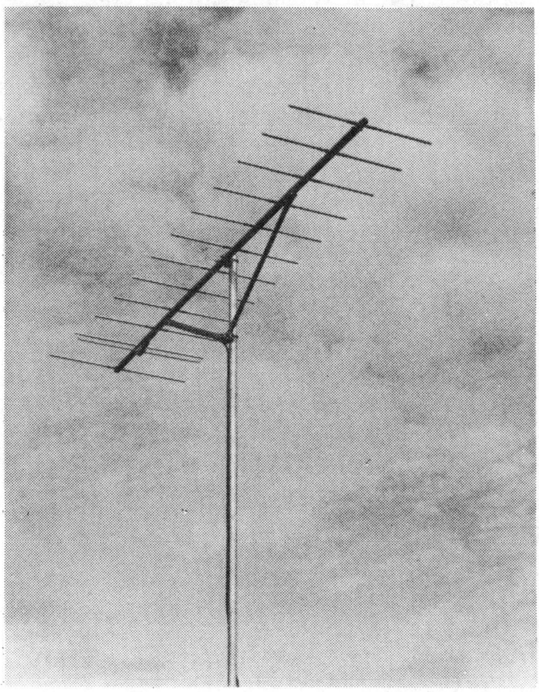

Courtesy SITCO Antennas

Fig. 2-2. SITCO CA 12-1 single-channel antenna.

The Winegard Chromstar line of single-channel antennas trades gain for increased bandwidth by using the log-periodic technique (explained in the next section). The driven element(s) in these antennas consists of a four-element log-periodic section instead of the usual folded dipole. This results in about 10 dB gain and 45° beamwidth for each of the nine-element low-band models, the CH-2002 to CH-2006. The ten-element high-band models, the CH-2007 to CH-2013, have much higher gain (11–13½ dB) and narrower beamwidth (38°–33°) because their drivers are peaked for higher gain and less percentage bandwidth, and the element spacing is greater. Except for element lengths, these antennas are identical in appearance to their multichannel counterparts shown later in Fig. 2-4.

Single-channel antennas are used in either of two places: at the head (antenna) end of a catv system and for homes in places so remote that only one or two channels are receivable. Catv systems use super-rugged (and commensurately expensive) antennas like the SITCO models, which have solid-rod elements and are built to withstand extreme wind and ice conditions. The outputs of a number of these antennas are individually amplified and combined with special devices to produce an equal-level multichannel signal.

Remote residences are the ideal application for quality consumer-grade single-channel antennas like the Winegard Chromstar line. It is ridiculous to use a broadband antenna to receive only one or two channels when a high-performance single-channel antenna will do a far better job for the money. The Weak-Signal Antenna Data table later in this chapter shows how single-channel antennas compare to single-band and multiband vhf antennas of similar size.

SINGLE-BAND ANTENNAS

The single-band antenna is a compromise between single-channel and all-channel antennas. Single-band vhf antennas cover either the vhf low band (vhf-L) or the vhf high band (vhf-H). These antennas are ideal for areas where several stations are available, but all are on either the vhf-L or vhf-H channels. More gain and better patterns are available in these circumstances by using a single-band antenna than by using an all-vhf channel antenna of similar size.

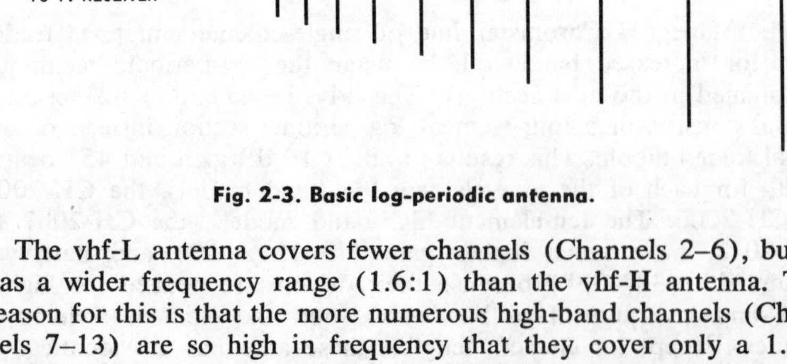

Fig. 2-3. Basic log-periodic antenna.

The vhf-L antenna covers fewer channels (Channels 2–6), but it has a wider frequency range (1.6:1) than the vhf-H antenna. The reason for this is that the more numerous high-band channels (Channels 7–13) are so high in frequency that they cover only a 1.2:1 frequency span. This narrower bandwidth requirement allows de-

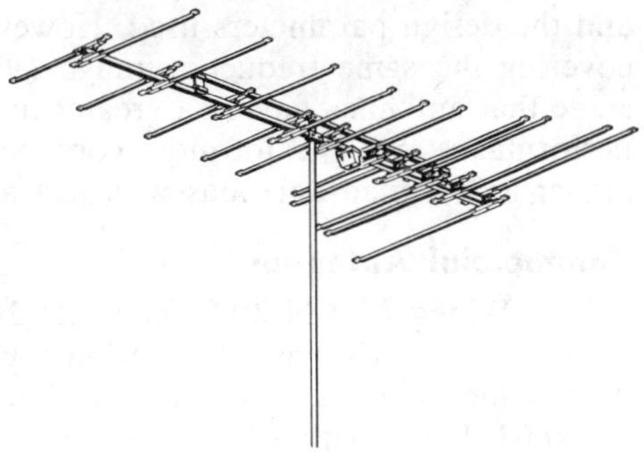

(*A*) *CH-2026.*

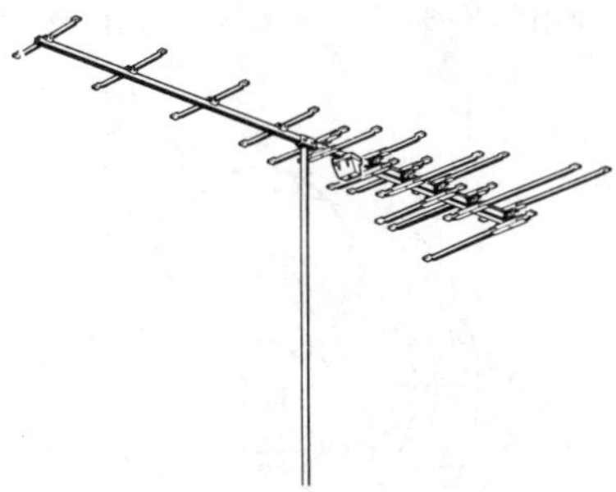

(*B*) *CH-2073.*

Courtesy Winegard Co.

Fig. 2-4. Winegard CH-2026 and CH-2073 single-channel antennas.

signing the vhf-H antenna for higher gain and narrower beamwidth than a vhf-L antenna of similar size.

The Log-Periodic Principle

All high-performance single-band antennas now achieve the required bandwidth by using a driven-element array that is a variation of the log-periodic antenna. The log-periodic antenna has a number of dipole elements connected with a transposed feed line as shown in Fig. 2-3. The longest element is resonant a little below the lower edge of the frequency band covered by the antenna; the shortest element is resonant above the upper edge of the frequency band. The spacing between adjacent elements, and the lengths of the elements, vary in a manner that results in the antenna having the same characteristics (i.e., gain, polar patterns, impedance, etc.) at every frequency within the frequency band for which the antenna is designed. In theory, the antenna bandwidth obtainable is unlimited; in practice, frequency ranges of 10:1 are feasible. The number of elements in the antenna depends on the frequency range to be covered and the design parameters used. However, when comparing antennas covering the same frequency range (vhf in this case), you can assume that antennas having a greater number of driven elements offer performance that is a lot more consistent with frequency (as well as higher gain), than antennas with just a few driven elements.

Commercial Antennas

The Winegard CH-2026 and CH-2073 antennas (Fig. 2-4) are high-quality single-band log-periodic yagis suitable for use on residences and matv systems. The CH-2026 provides 7–8½ dB gain over the vhf-L band, with a beamwidth varying from 63° at Channel 2 to 50° at Channel 6. Because of the smaller frequency-coverage ratio and wider element spacing, the CH-2073 vhf-H antenna has higher gain (8½–13½ dB) and beamwidths varying from 52° at Chan-

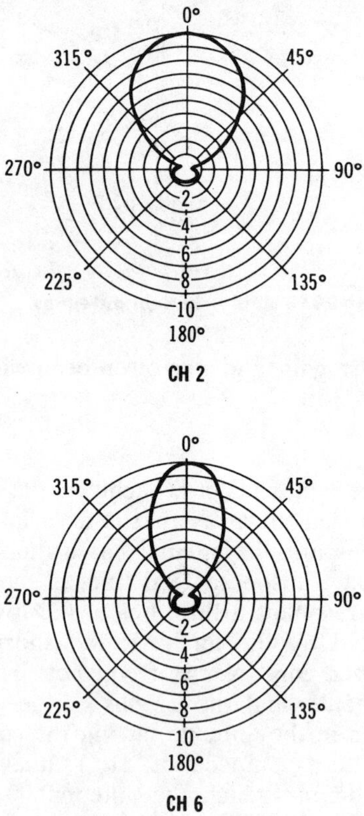

(*A*) *CH-2026 patterns.*

Fig. 2-5. Polar patterns for Winegard

nel 7 to 38° at Channel 13. The front-to-back ratio for either antenna is 20 dB or more on any channel.

Manufacturer's polar patterns for the CH-2026 and CH-2073 are shown in Fig. 2-5. The Weak-Signal Antenna Data table later in this chapter shows how these single-band antennas compare to single-channel and multiband vhf antennas of similar size.

MULTIBAND ANTENNAS

Multiband vhf antennas cover both the high and low vhf tv bands, and thus operate on all vhf channels (2–13). The multiband antenna is the most commonly used vhf antenna, and it is the only kind practical for residences where several high-band and low-band vhf stations are received.

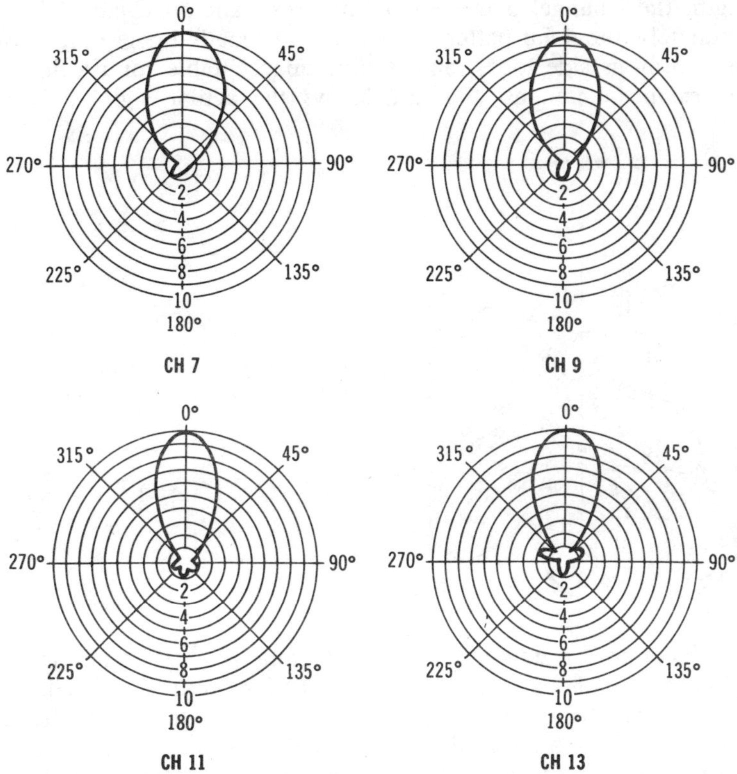

Courtesy Winegard Co.

(B) CH-2073 patterns.

CH-2026 and CH-2073 antennas.

The frequency ratio of Channel 13 to Channel 2 is 4:1. A pure log-periodic antenna can readily be built to handle this range, but it would have relatively low gain per channel for its size, and it would be quite expensive. Fortunately, a technique known as *multimoding,* or *bimoding* allows a relatively-high-gain antenna with a 1.6:1 frequency span to work well on all vhf channels. This results in great savings in antenna size, weight, and cost.

Multimoding Techniques

The principle behind multimode operation is the fact that the vhf-H channel frequencies are three times higher than the vhf-L channel frequencies. This means that an antenna element a half-wavelength long on Channel 3 (for example) is 1½ wavelengths long on Channel 9. Since resonance occurs in multiple of a half-wavelength, the Channel 3 element is also resonant on Channel 9. Unfortunately, the polar pattern of a 1½-wavelength dipole (Fig. 1-8A) has enormous side lobes, and is thus unacceptable for use in high-performance antennas. Fig. 2-6 shows the actual high-band polar

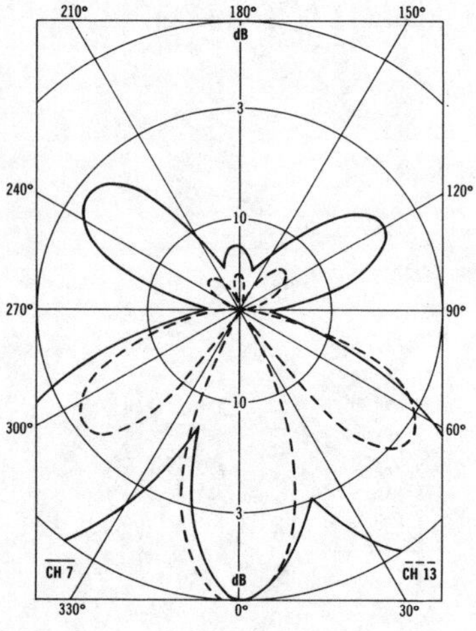

Fig. 2-6. High-band polar pattern of vhf antenna with proper multimoding.

patterns of an alleged multiband vhf antenna that had no multimoding techniques. Although its performance and pattern are excellent on the vhf low band, high-band performance is unacceptable. The enormous side and back lobes are not only indicative of susceptibility to ghost pickup, but also of low forward gain. In fact, the Channel 7 pattern shows greater signal pickup from the side lobes than from the front lobe!

Multimoding techniques alter the antenna elements in a way that makes a 1½-wavelength element act like a one-wavelength element,

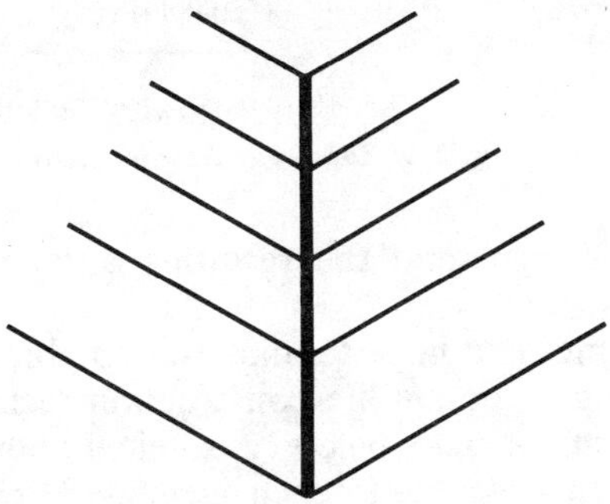

Fig. 2-7. Vee'd antenna.

which has a pattern with a large narrow main lobe but no side lobes. Thus, a log-periodic antenna designed for good performance on the vhf low band can be made to produce high gain and a good polar pattern on the vhf high band. The three most popular ways of doing this use either vee'd elements, parasitic elements, or extension stubs.

The simplest multimoding technique is vee'd elements (Fig. 2-7). The elements are swept forward to distort the six-lobe pattern of the 1½-wavelength element in a manner that concentrates a larger amount of the total antenna gain in the forward center lobe (Fig. 2-8). However, the ideal angle for any given element is different for each channel on which that element operates. Since the amount of forward sweep must be a compromise, side-lobe suppression is excellent on some high-band channels and only fair on others. Another fault is that low-band gain is reduced somewhat by the forward sweep. This technique does work quite well overall with the

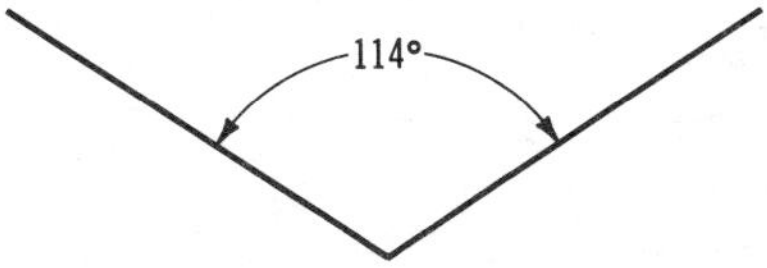

(A) Element angle.

(B) Polar pattern.

Fig. 2-8. Swept-forward dipole.

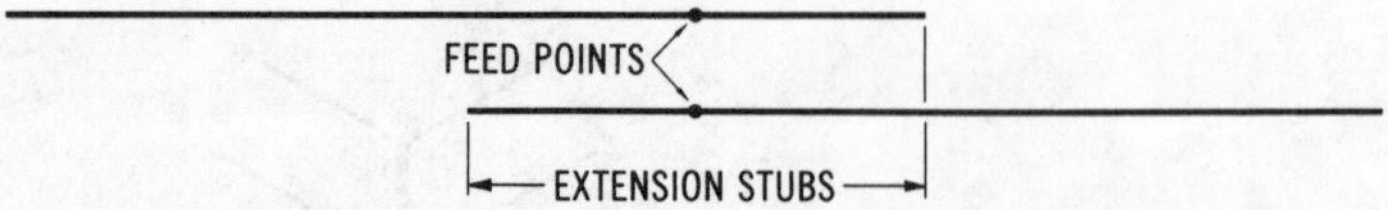

Fig. 2-9. Edgewise view of element with extension stubs.

right design, and the resulting antenna is very light and attractive for its size.

A multimoding technique adaptable to log-periodic antennas employing a compatible construction technique (double booms or feed lines above and below a single boom) is that of extension stubs. With this technique each element is extended at the feeder lines to produce the equivalent of a half-wave dipole on the vhf high band. These extensions (shown in Fig. 2-9) have too high an impedance on the vhf low band to affect low-band operation, but on the high band they become the dominant elements and cause the antenna to perform on the high band very much like it does on the low band. The high-band polar patterns are generally cleaner (fewer and smaller minor lobes) than those produced by any of the other multimoding techniques. However, this similarity between vhf-L and vhf-H operation means the vhf-H patterns are broader (wider beamwidth) than those produced by the other multimoding techniques, and that high-band gain is correspondingly lower.

The most popular multimoding technique uses parasitic elements. These are high-band half-wave dipoles installed near the low-band driven elements (Fig. 2-10). The parasitic elements buck out the center currents of the driven elements, allowing the driven elements to work like full-wave dipoles on the vhf high band. This results in higher high-band gain than is possible with the other techniques already discussed, but the side-lobe suppression is not as good as the extension-stub technique. When you look at the high-band polar patterns of the antennas in this chapter, and at the vhf-H patterns of

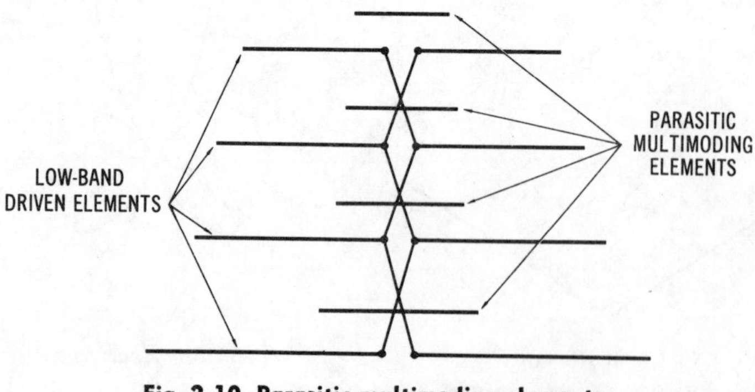

Fig. 2-10. Parasitic multimoding elements.

the vhf-uhf antennas in Chapter 4, you will see some evidence of the classic six-lobe patterns in all antennas multimoded by parasitics, vee'ing, or the proprietary technique discussed next.

Jerrold Electronics uses a unique technique in their VIP series of vhf antennas, and their VU series of vhf-uhf antennas. The technique involves a combination of element feed-line impedance and construction, element spacing, and special element configurations. The net result is good polar patterns *and* excellent high-band gain. This is the only nonstandard multimoding technique that really works well.

Multimode Directors

Directors can also be made to function in both vhf bands. The multimode director shown in Fig. 2-11 is almost universally used on large antennas. Depending on the manufacturer, it may be referred to as a dual-mode, bimode, multimode, or high/low director. In this book the term *high/low* will be used because it is the most descriptive.

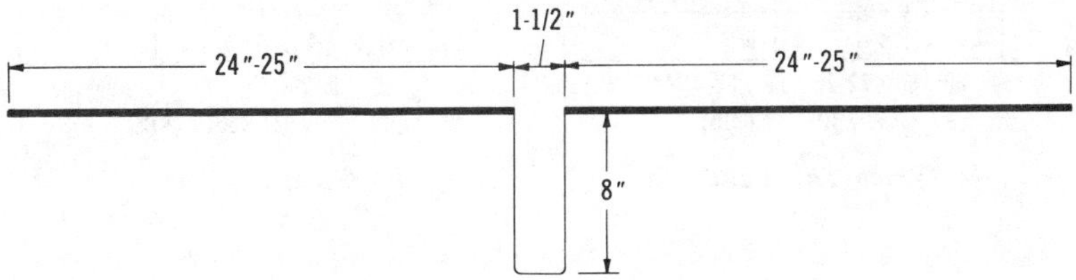

Fig. 2-11. High/low vhf director.

A high/low director is actually two collinear high-band directors connected with a hairpin of aluminum wire. The hairpin's dimensions are such that it appears as an open circuit at the high band, allowing the elements to function as a pair of collinear half-wave directors resonant at 216 MHz (Channel 13). At low-band frequencies the hairpin acts as a length of wire converting the two short elements to one long element resonant at 88 MHz (Channel 6). Thus each high/low director acts as one vhf-L director and two vhf-H directors, and is so considered in the element counts in this chapter and Chapter 4.

FM Reception

Some vhf antennas are designed so the low-band coverage extends from 54 to 108 MHz, instead of ending at 88 MHz. This allows a single antenna to provide signal to an fm receiver or tuner, as well as to the tv set. However, fm gain comparable to vhf-L gain is the exception rather than the rule. The low-band gain of the typical small tv antenna falls off rapidly past Channel 6, as shown in Fig. 2-12A. A small antenna with good fm coverage may have a gain curve like that in Fig. 2-12B.

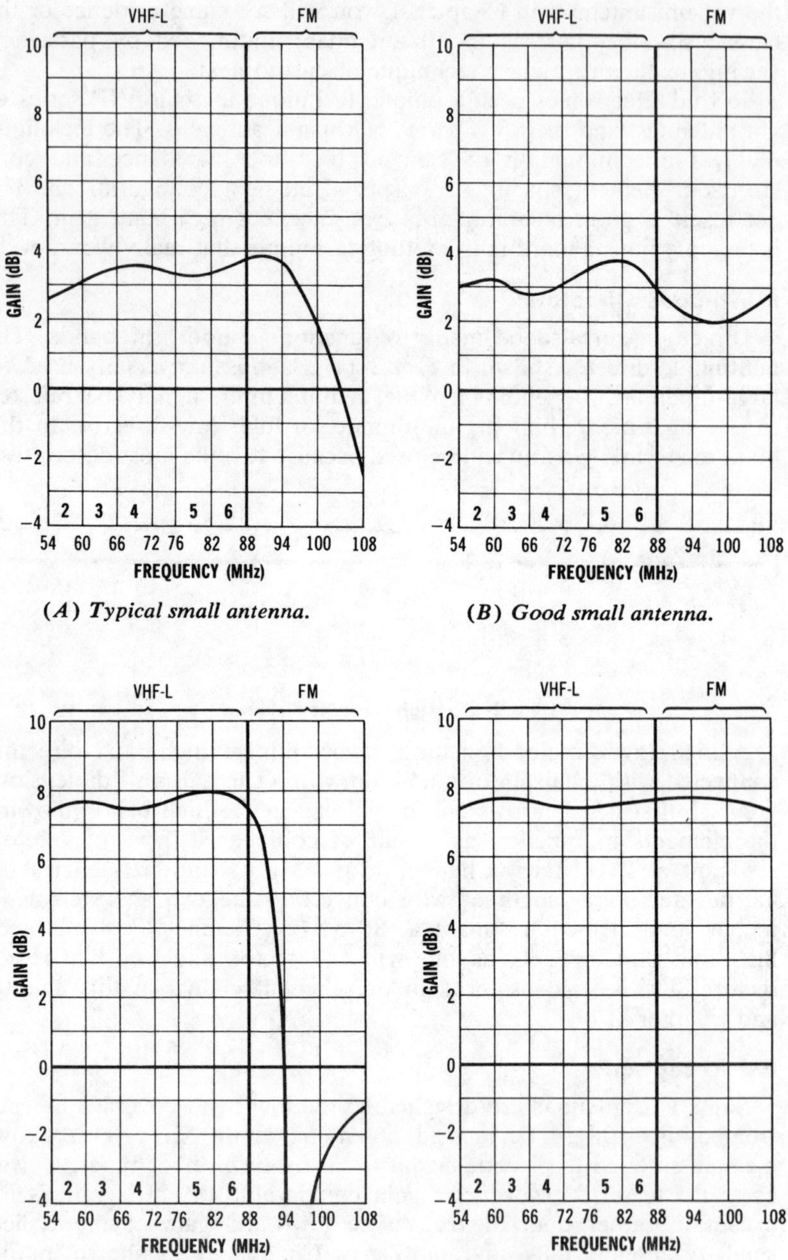

(*A*) *Typical small antenna.*

(*B*) *Good small antenna.*

(*C*) *Full-length directors.*

(*D*) *Shortened directors.*

Fig. 2-12. The fm gain of vhf tv antennas.

The difference in some medium and most large antennas is much more severe. Because of the use of high/low directors, the gain of these antennas drops rapidly past the director design frequency (88 MHz). This results in fm gain that is 10–15 dB below that of Channel 6 (Fig. 2-12C). This characteristic is very useful when nearby fm stations interfere with Channel 6 reception. However, good fm gain is obtainable with antennas of this type by shortening the high/low elements. This raises the resonant frequency to 108 MHz to boost the gain in the fm region (Fig. 2-12D). Most manufacturers either provide score marks on the high/low elements (permitting their ends to be snapped off) or specify the length to be cut off each element. Unfortunately, the shortening greatly reduces the effectiveness of the directors at tv frequencies, so both the vhf-L and vhf-H gain is lowered ½ dB or more on some channels (there is no free lunch).

COMMERCIAL MULTIBAND ANTENNAS

The remainder of this chapter features a number of commercially available multiband vhf antennas that have earned the author's approval for their fine performance and noteworthy features. Most of the small- to medium-sized antennas were personally measured and evaluated by the author. Since a few antennas were not personally tested, the recommendation here is based on the measured performance of another antenna in the same line and the detailed performance data from the manufacturers.

As much information as possible is given as to why these particular antennas perform well, and why they are suitable to a particular application. As this was written a potential purchaser could select the antenna that best fit his or her needs from these pages. Future purchasers must learn to recognize the characteristics of a fine antenna because most of the models shown here will eventually be discontinued and replaced by others. Manufacturers are forced to move on to designs that are cheaper to manufacture or easier to retail in order to remain competitive in the face of rising materials and labor costs. Sometimes the new models are better than the ones replaced, sometimes they are not. Because of a lack of books like this one, tv antennas have traditionally been bought on the basis of price versus number of elements. Since the best-designed antennas in the world could therefore gather dust in a warehouse because they cost 10 percent more than similar-sized garbage, manufacturers are often forced to scrap some good designs to stay in business.

Element count is another matter that should be discussed before comparing multiband antennas. Some manufacturers count every half-wavelength rod as an additional element, regardless of its function. This gives small antennas multimoded by parasitics a higher element count than larger antennas multimoded by vee'ing or Jerrold's technique. Some manufacturers even count every noncontinuous rod as an element. This means that each half-wavelength centerfed element is counted as two elements, giving a six-element antenna, such as the RMS Electronics DJR-6, an element count of 12! The gist of this is that you should ignore manufacturers' element counts and make your own (as did the author) when comparing antennas. Count the number of half-wavelength elements functional on the low band (reflector, driven elements, low-band directors) and the number of elements functional on the high band (multimoded driven elements, high-band directors) for each vhf antenna you compare.

Strong-Signal Antennas

Within 10–12 miles of a full-power vhf station's antenna, insufficient signal strength is no problem. Rather, too much signal may be the problem, since not all tv receivers can handle several channels in excess of 50,000 microvolts apiece. These output levels are possible from a medium-sized vhf antenna located within 5 miles of the transmitting antenna. The problem in metropolitan markets where such signal levels are common is with reflected signals (ghosts). The antennas needed in these areas are low-gain antennas with excellent patterns and high front-to-back ratios to minimize ghost pickup. High gain and good patterns go together, however, so most two- and three-element antennas do not have patterns suitable for metropolitan areas (or anywhere else).

Large high-gain antennas are usually required to provide the necessary narrow beamwidth and high front-to-back ratios needed to reject ghost signals. Their output can be attenuated to a suitable level, but large antennas are expensive, unsightly, and their size and weight make them difficult to mount securely. Winegard offers an excellent alternative in their CH-4210 (Fig. 2-13A). This is a low-gain antenna with the narrow beamwidth and high front-to-back ratios (see Fig. 2-13B and 2-13C) of a good high-gain antenna. This antenna uses just two driven elements connected with a special phasing network, a reflector, and three high-band directors to produce good polar patterns. The 75-ohm output impedance of this network permits a direct match to coaxial cable, so transmission-line pickup cannot occur and degrade the signal.

The gain curve of the CH-4210 is highly tilted, a characteristic useful when the cable run is extremely long. According to the manufacturer's data, the gain varies from −3 dB at Channel 2 to +5 dB at Channel 13. This "tilt" will cancel the reverse tilt in 350 feet of RG59 foam coax (Belden 9275) or 440 feet of RG6 foam coax

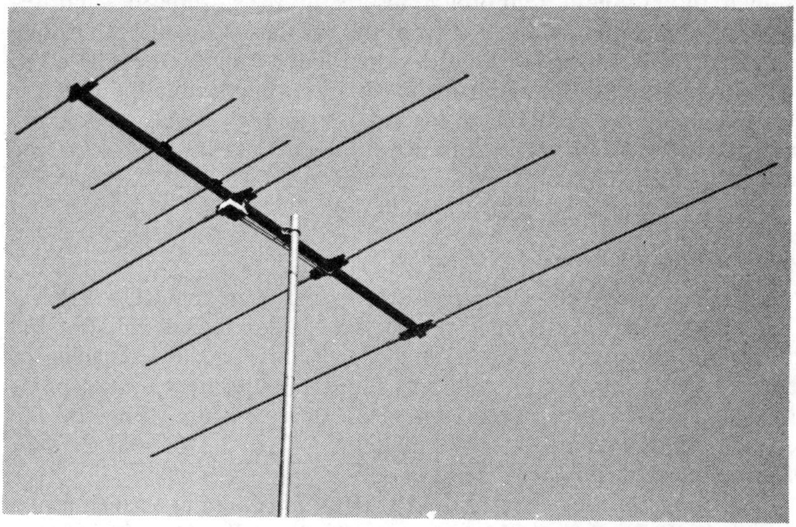

Courtesy Sony Corp. of America

(*A*) *Photograph of antenna.*

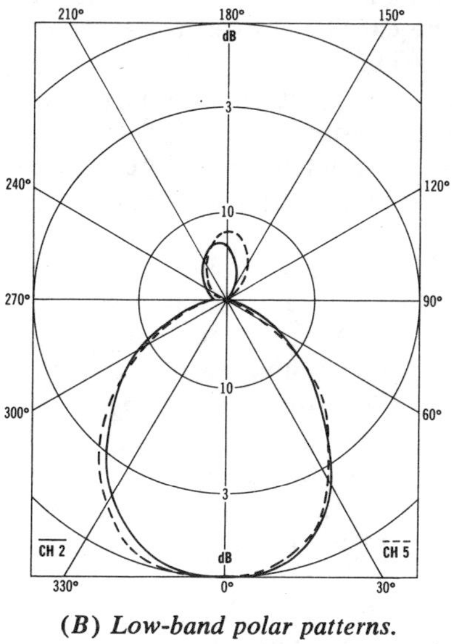

(*B*) *Low-band polar patterns.*

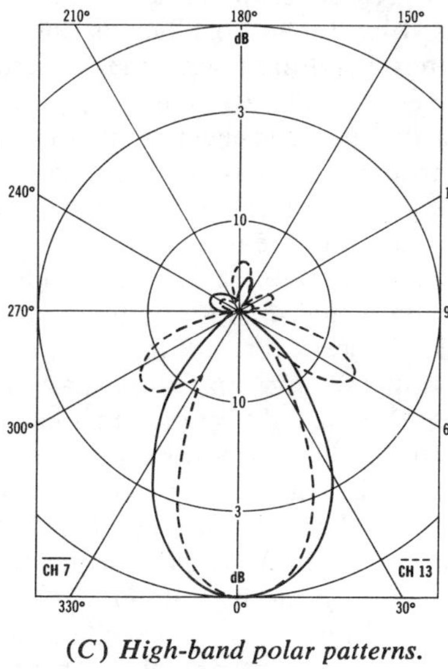

(*C*) *High-band polar patterns.*

Fig. 2-13. Winegard CH-4210 antenna and polar patterns.

(Belden 9283), resulting in a flat antenna/transmission-line combination.

The physical characteristics of the CH-4210 are also excellent. The CH-4210 has $\frac{7}{16}$-inch-diameter elements (instead of the usual

⅜-inch tubing) with reinforcing inserts at the insulators, and reinforcing sleeves extending out from the insulators on the three long elements. This unusually rugged construction makes the CH-4210 one of the strongest antennas on the market, highly qualifying it for installation in metropolitan areas subject to high winds and icing, such as those in the northeast and Great Lakes sections of the country. Lastly, the neat appearance of the CH-4210 will not result in an eyesore atop your dream house!

Medium-Signal Antennas

For locations 10–30 miles from the transmitter, the RCA 3BG09 (Fig. 2-14A) is an excellent choice. This antenna has a high-band director, three driven elements, and a reflector. The wide spacing of these elements and a combination of multimoding techniques (parasitic element for the first driven element, extension stubs for the second) result in the best polar patterns (Figs. 2-14B and 2-14C) of all antennas tested in its price class.

According to the manufacturer's data, this antenna averages a smooth 2-dB gain on the low band, and a 4–6 dB on the high band. However, the 3BG09 (and its vhf-uhf equivalents in Chapter 4) had greater Channel 2 output than many of the larger antennas tested by the author. This is due to the excellent vertical-plane polar pattern resulting from wide element spacing.

The 3BG09 has sufficient gain for use in the near suburbs, and can drive a high-level signal-distribution system (see Chapter 6) in metropolitan areas without the need of amplification. Moreover, its fine polar patterns will minimize pickup of the ghosts so prevalent in metropolitan areas.

The Antennacraft Mark-8 (Fig. 2-15A) is a compact antenna having a high-band director and four driven elements. It uses parasitic elements for multimoding, so gain is favored in the design. The gain of this antenna is unusually high for a relatively small (50-inch boom) and lightweight antenna, so much so that its average high-band gain was in the same category as some larger and more costly antennas. This performance is evidenced by the narrow beamwidth of the high-band patterns (Fig. 2-15C). A reasonable range expectation for a Grade A picture is about 30 miles on the low band, and 40 miles on the high band.

The Mark-8 is an excellent antenna for the suburbs, especially where a large antenna would detract from the appearance of a home. This antenna is also ideal when a high-performance antenna is needed for a mobile home or recreational vehicle. Most of the "special' antennas sold for these applications have high prices and poor performance; the Mark-8 on a conventional mast is a superior alternative. With the aid of a small screwdriver or knife blade the elements

Courtesy Sony Corp. of America

(*A*) *Photograph of antenna.*

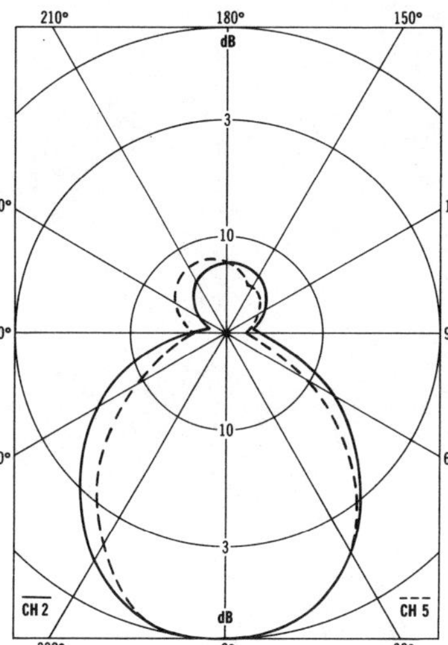

(*B*) *Low-band polar patterns.*

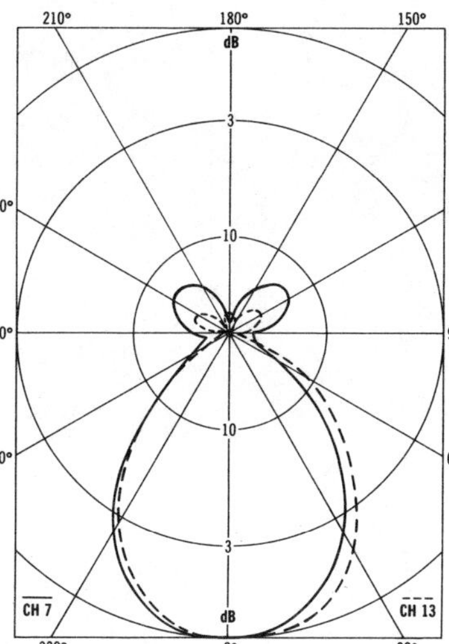

(*C*) *High-band polar patterns.*

Fig. 2-14. RCA 3BG09 antenna and polar patterns.

can be refolded flat against the boom, and the mast dropped to a traveling position in a matter of minutes.

Moderate-sized antennas (70- to 80-inch booms) give excellent results in most cases up to 40 miles from the transmitter. The RMS

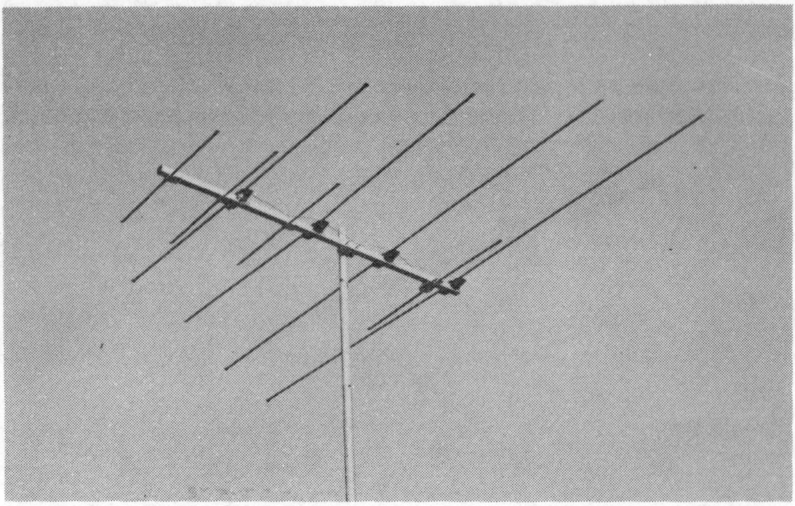

Courtesy Sony Corp. of America

(A) Photograph of antenna.

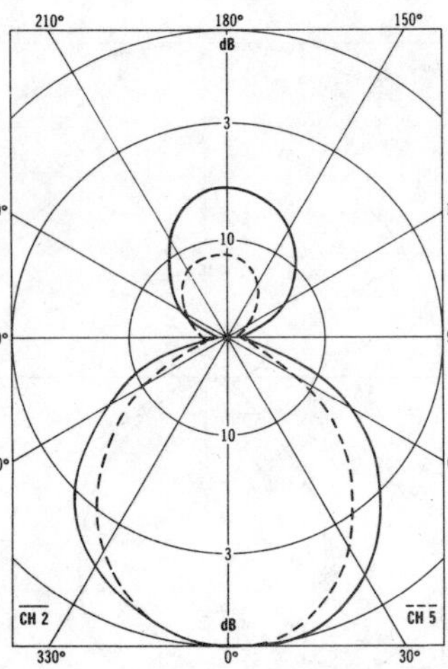

(B) Low-band polar patterns.

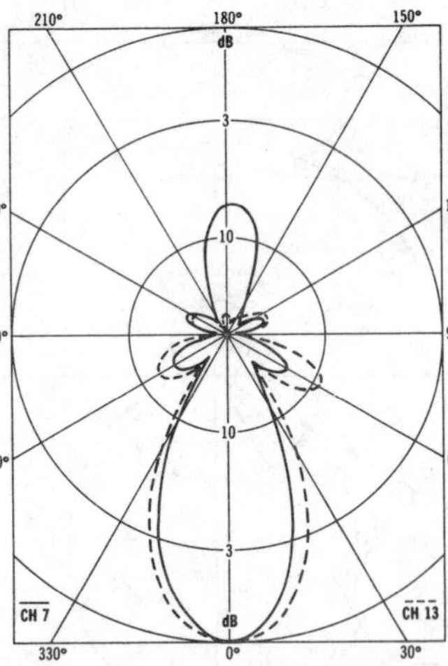

(C) High-band polar patterns.

Fig. 2-15. Antennacraft Mark-8 antenna and polar patterns.

Electronics DJR-6 (Fig. 2-16A) is a neat-looking lightweight antenna that is multimoded by vee'ing the elements. There are six driven elements, but the last one is shorted with a wire hairpin, so it acts as a reflector on the low band. Its outstanding characteristic is excellent low-band polar patterns. As Fig. 2-16B shows, the DJR-6

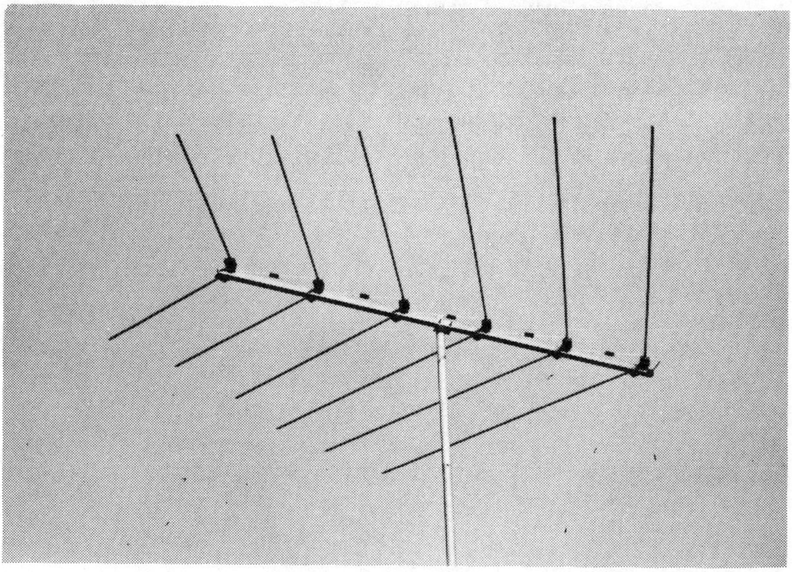

Courtesy Sony Corp. of America

(A) Photograph of antenna.

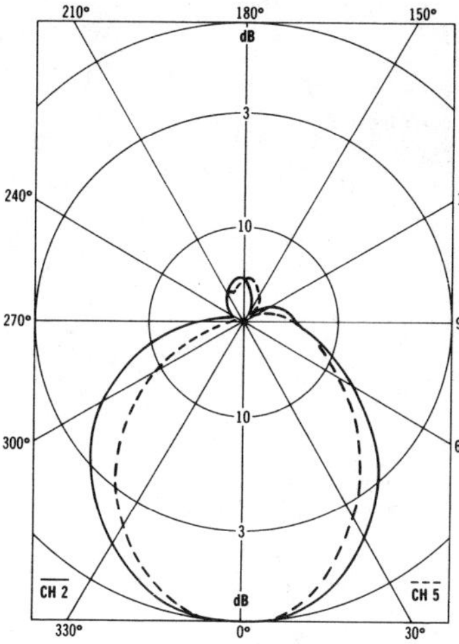

(B) Low-band polar patterns.

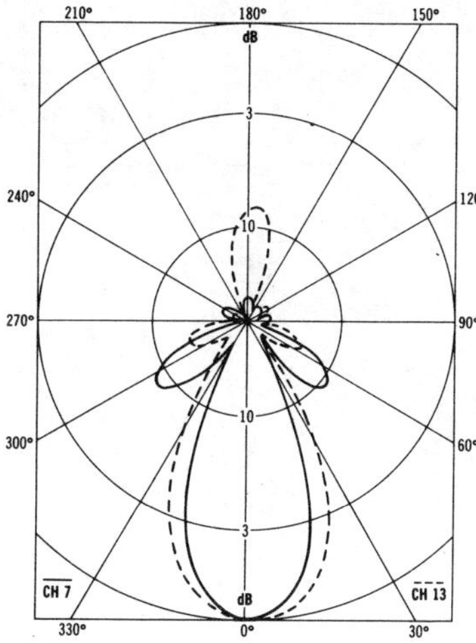

(C) High-band polar patterns.

Fig. 2-16. RMS Electronics DJR-6 antenna and polar patterns.

has high front-to-back ratio and deep side nulls. The side nulls are important in areas where fm broadcast interference occurs on Channel 6; the fm signal can be eliminated by reorienting the antenna to drop the fm interference into a side null.

The Antenna Corp. of America AC-511 (Fig. 2-17A) follows the Winegard/A.C.A. philosophy of packing a lot of elements on a small boom to achieve maximum gain for the antenna length. The AC-511 has four driven elements, a reflector, a high/low director, and three high-band-only directors on a 71-inch boom! Despite the fact that parasitic techniques are used for multimoding, the high-

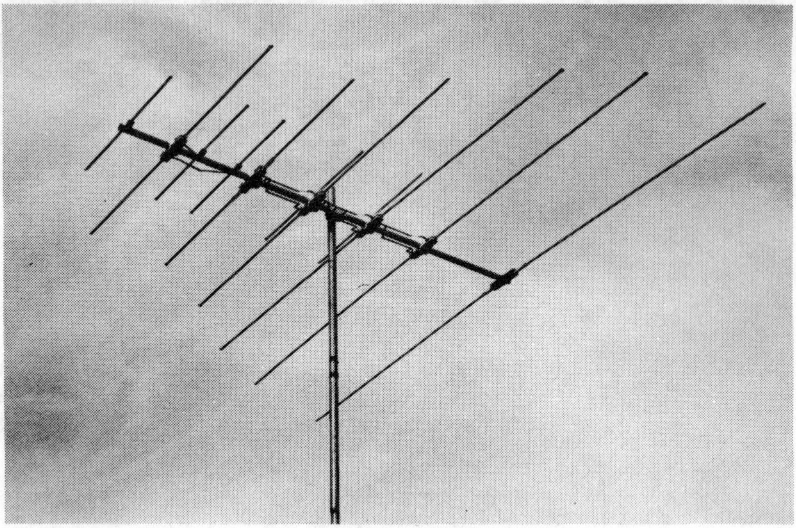

(A) Photograph of antenna.

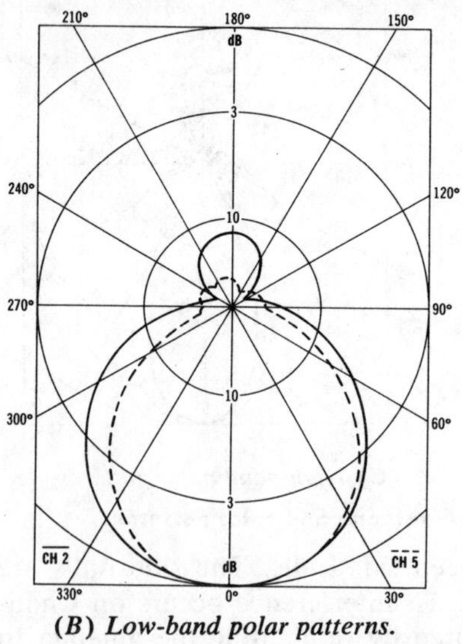

(B) Low-band polar patterns.

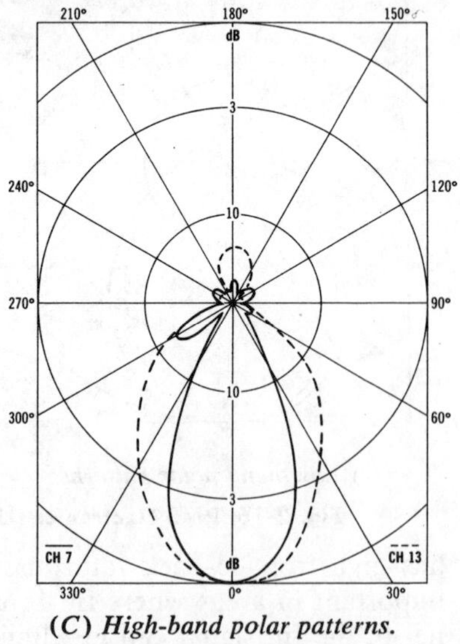

(C) High-band polar patterns.

Fig. 2-17. Antenna Corp. of America AC-511 and polar patterns.

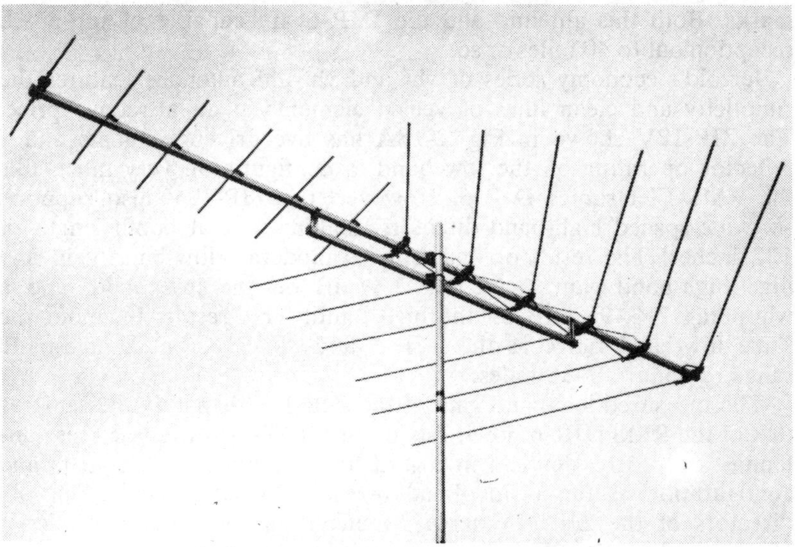

(A) Photograph of antenna.

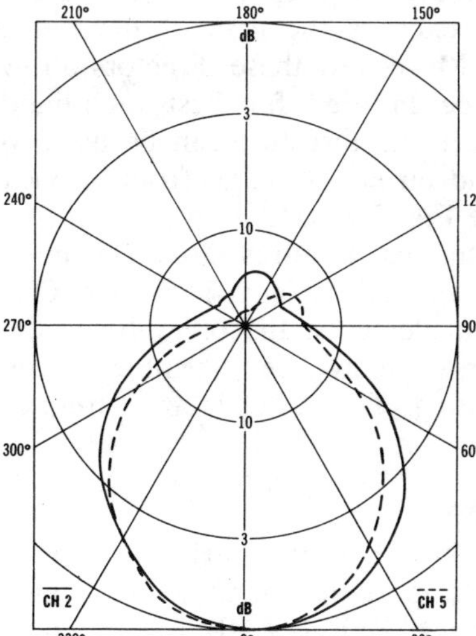

(B) Low-band polar patterns.

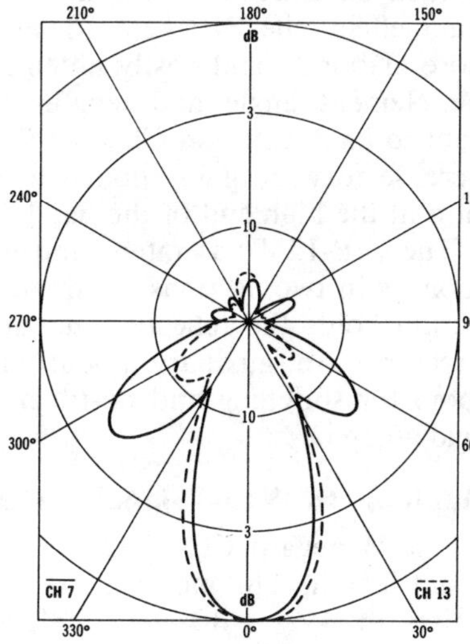

(C) High-band polar patterns.

Fig. 2-18. Jerrold ZIP-12V antenna and polar patterns.

band patterns (Fig. 2-17C) are very clean (i.e., small minor lobes). As shipped, the antenna is designed for minimum fm pickup. If fm reception is desired, it can be obtained at the cost of some loss of tv gain by snapping the ends off the high/low director at the scored marks. Both this antenna and the DJR-6 are capable of top-notch reception out to 40 miles or so.

Jerrold's economy series of vhf and vhf-uhf antennas features the simplicity and clean lines of vee'ed elements at an attractive price. The ZIP-12V shown in Fig. 2-18A has five driven elements and a reflector operating on the low band, a configuration very much like the RMS Electronics DJR-6. However, the ZIP-12V also supports six wide-spaced high-band directors, making a total boom length of 122 inches! The result of this setup is moderate low-band gain and high high-band gain, averaging 3½ dB on the low band, and a whopping 7½–9½ dB on the high band! This results in about the same low-band range as the DJR-6 and AC-511, but a high-band range of about 50–55 miles.

The measured low-band gain of the ZIP-12V is nearly the same as that of the RMS DJR-6 previously discussed. This is because these antennas are nearly identical in regard to the number of elements and configuration as far as low-band operation is concerned. (The six directors of the ZIP-12V have no effect on low-band operation.) However, the ZIP-12V is vastly superior from Channel 9 and higher. In fact, on Channels 11 and 13, where the six directors have maximum effect, the ZIP-12V equals or exceeds the gain of the much more elaborate and costly antennas. Moreover, these directors allow the element angle and lengths to be tailored for best high-band front-to-back ratio on Channel 7, since the directors can be used to increase forward gain enough to yield an equally-high front-to-back ratio at the high end of the band (see Fig. 2-18C).

The ZIP-12V has interesting mechanical features also. The main boom is in two sections, so the carton length is only 68 inches. One section holds the reflector and driven elements, the other holds the directors. An auxiliary boom mounts below the assembled main boom for stiffening and no-tilt mounting. The result is an extremely rigid antenna.

Medium- to Weak-Signal Antennas

The Winegard CH-4052 (Fig. 2-19A) has very high gain for its boom length. The measured results support the manufacturer's gain claims of 4½–6 dB for the low band, and 8–9 dB for the high band. This will provide sufficient output for a good picture at reception sites 50–55 miles from the transmitter over average terrain. Although only 75 inches long, the CH-4052 has a relatively high element count: six driven elements (rearmost shorted to act as a reflector), three exclusively high-band directors (one of which is the Winegard broadband double-director described at the end of Chapter 1), and a high/low director (which acts as one low-band director and two additional high-band directors). Multimoding is by means of extension stubs, so the very clean high-band patterns shown in Fig. 2-19C are the result.

The CH-4052 is quite heavy (6 lbs, or 2.7 kg) because of its many elements and its very rugged construction. Like the CH-4210

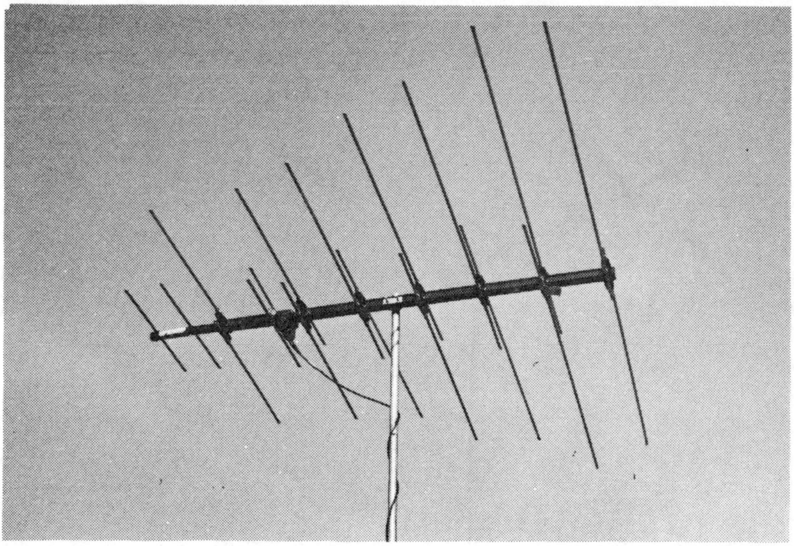

Courtesy Sony Corp. of America

(*A*) *Photograph of antenna.*

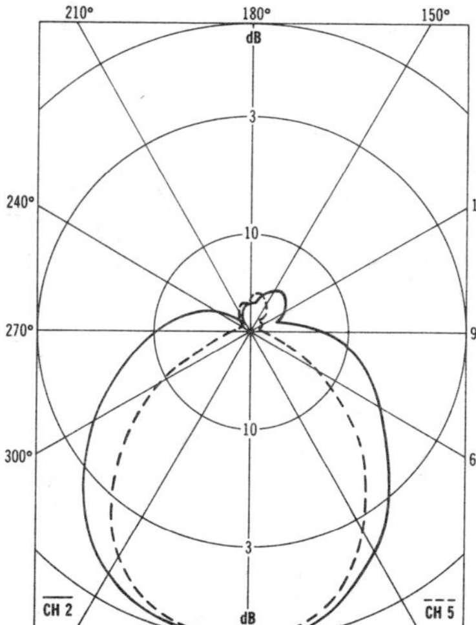

(*B*) *Low-band polar patterns.*

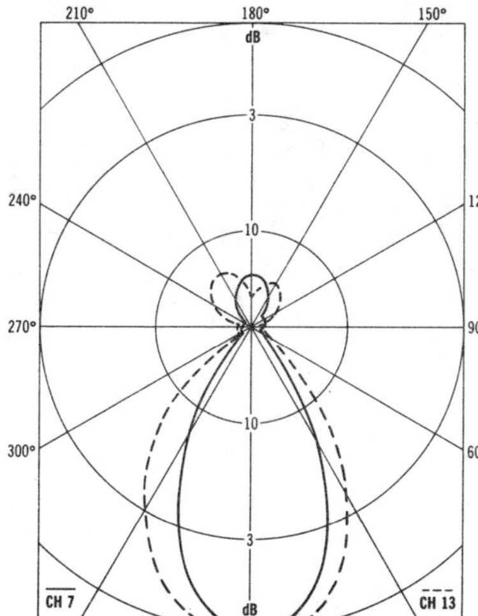

(*C*) *High-band polar patterns.*

Fig. 2-19. Winegard CH-4052 antenna and polar patterns.

described earlier, the CH-4052 has large-diameter (7/16-inch) elements with supportive insulators and reinforcing sleeves extending out from the insulators, making it one of the strongest antennas available on the consumer market. This combination of rugged con-

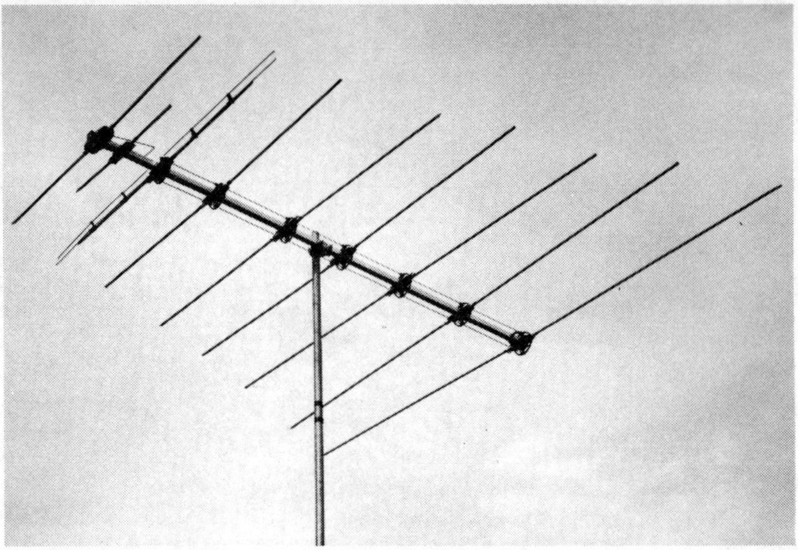

(*A*) *Photograph of antenna.*

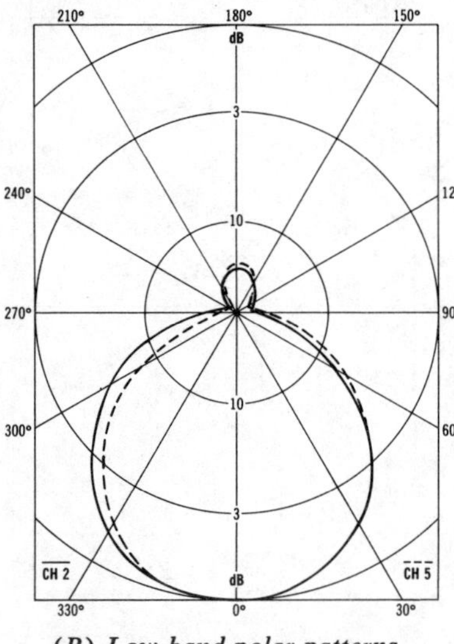

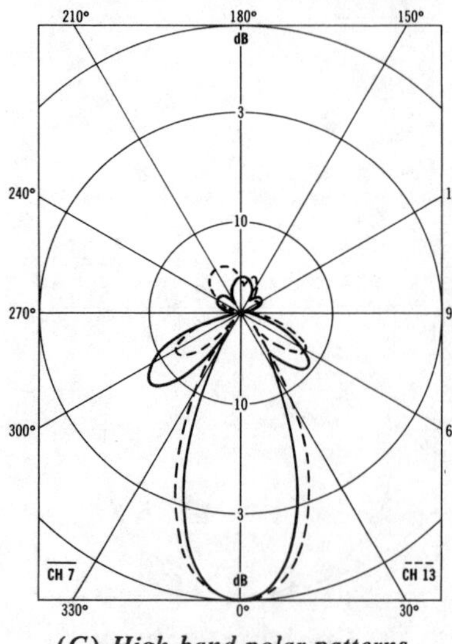

(*B*) *Low-band polar patterns.* (*C*) *High-band polar patterns.*

Fig. 2-20. Jerrold VIP-303 antenna and polar patterns.

struction and gain makes it a top choice for medium- to weak-signal areas subject to high winds and icing, and for states in the hurricane belt.

A weatherproof housing on the boom encloses a cartridge that offers a choice of 75- or 300-ohm output impedance. This housing will also accept fm and CB traps, or one of the AC series preamplifier cartridges described in Chapter 6. Instructions supplied with this antenna show how to make the connections for the desired impedance, and how to install the various devices in the housing.

A high-gain antenna employing many unusual techniques is the Jerrold VIP-303 shown in Fig. 2-20A. This antenna uses Jerrold's unique multimoding system, which is actually a combination of design approach and construction technique. First is the 450-ohm over-and-under transmission line that connects its seven driven elements, the rearmost of which is shorted to act as a reflector. Next is the "trombone" first driven element; capacitive coupling to the wire hairpin makes the first element seem much longer at Channel 7 than it does at Channel 6. This reduces the Channel 7 side lobes without impairing reception on Channel 6. This technique is also responsible for the excellent fm performance possible from this antenna. As shipped, however, the VIP-303 inhibits fm reception by 12 dB or more. When the high/low director is shortened (at some cost in tv gain) by breaking off its ends at scored marks, its resonant frequency is increased to 108 MHz and full fm gain results. So this and all larger VIP series antennas offer a choice of fm rejection or quality fm reception.

The result of these practices is a claimed 4½–5 dB low-band gain, and 8½–9 dB high-band gain. The author's measurements support these claims. In fact, the VIP-303 had the highest average measured gain of the antennas tested. This is partly due to its relatively long boom length for the number of elements. The range estimate is therefore around 55 miles over average terrain for a Grade A picture.

The front-to-back ratios are good at all frequencies, as Figs. 2-20B and 2-20C show, but the outstanding pattern characteristic of the VIP-303 is excellent low-band vertical-plane polar pattern. This characteristic, which is generally related to boom length, makes the antenna insensitive to ground (or roof) reflections and to interference coming from below or above the plane of the antenna.

Weak-Signal Antennas

A really high-gain antenna is needed when the reception site is 60 or more miles from the transmitter, or when the intervening terrain blocks the signal. Antennas that can do the job at these distances are extremely long and often costly. Their great size entails more than

Table 2-1. Weak-Signal Antenna Data

Model	List Price ($)	Boom Length (in)	Boom Type	Wt (kg)	Directors VHF-L	Directors VHF-H	Driven Elements	Multimoding Technique	Gain (dB) VHF-L	Gain (dB) VHF-H	Beamwidth (°) VHF-L	Beamwidth (°) VHF-H	Front-to-Back Ratio (dB) VHF-L	Front-to-Back Ratio (dB) VHF-H
Winegard CH-4053	66	109	S	3.9	3	11	5	Ext. stubs	6.4	9.4	63	43	19	20
Jerrold VIP-304	67	110	S	—	3	6	7	Complex	5.5	9.8	68	37	16	19
Winegard X-220	36	117	S	—	2	8	6	Parasitics	5.6	10.5	61	41	20	27
Jerrold VIP-305	80	121	D	—	3	6	9	Complex	6.3	11.0	67	36	18	21
RCA 3BG27	87	164	D	5.4	4	16	6	Para. & stubs	6.3	9.2	62	38	20	23
Winegard CH-4054	85	157	D	5.8	5	17	6	Ext. stubs	6.9	10.4	66	38	19	19
Winegard CH-5200	116	200	D	6.5	5	18	9	Ext. stubs	7.2	11.9	62	36	20	20
Jerrold VIP-307	120	194	D	—	15	30	9	Complex	6.7	12.9	65	32	24	24
Winegard CH-2026	64	144	D	3.6	4	—	4	—	7.5	—	55	—	20	—
Winegard CH-2073	41	125	S	2.2	—	5	4	—	—	11	—	44	—	20
Winegard CH-2004	64	144	D	3.6	4	—	4	—	10	—	44	—	20	—
Winegard CH-2010	41	125	S	2.2	—	5	4	—	—	12	—	37	—	20

Wt = Weight Para. = Parasitics D = Double S = Single Ext. = Extension just a high price; size means high wind resistance and high weight. These factors demand solidly built supporting structures, such as heavy-duty, well-guyed masts or rugged towers. The antenna rotators for the heavier of these antennas must also be heavy-duty devices, as 150–200 inch heavy-duty antennas produce a tremendous counter-torque when ice loaded in gale winds.

Because of their unwieldiness and costliness, no weak-signal antennas were tested by the author. Instead, the recommendations in this section are based on the author's measurements of smaller models in the lines, and manufacturers' data known to be reliable.

The models approved for this section all had gain curves that were reasonably flat across the band and had no more than 1-dB variation across any channel. The polar patterns were all free from minor lobes to a degree deemed appropriate for the antenna size and price range.

Statistical data for the recommended weak-signal antennas are given in Table 2-1. In addition to the elements mentioned, each antenna also has one reflector or reflector equivalent (shorted driven element). The antennas are arranged roughly in order of increasing range. These range estimates represent high-quality reception at distances of about 55 miles at the top of the table to about 75 miles at the bottom. Far greater range is possible for any of these antennas if the antenna is mounted unusually high above favorable terrain, or if a lesser-quality picture is acceptable.

The Winegard Chromstar antennas (CH prefix) in Table 2-1 and Fig. 2-21 have the same construction and features as the CH-4052 mentioned in the previous section. On the other hand, Winegard's X-220 (Fig. 2-22) is a very lightweight antenna suitable for mild climates, hence its low price. Although only one step up in the line from the 4052, the CH-4053 is 34 inches longer. The additional driven element and two extra high/low directors give the 4053 a much narrower vertical-plane polar pattern. Therefore, the 4053 should be used instead of the short-boom 4052 in situations requiring a very rugged antenna when ground reflected or originated interference is a problem, or when maximum low-band gain is needed but the antenna is mounted less than 15 feet above the roof.

The RCA 3BG27 (Fig. 2-23A) is the outstanding performer at the high end of RCA's antenna line. The excellent rising gain characteristic (see Fig. 2-23B) suits it to driving long cables without the need of a tilt attenuator (Chap. 6). The polar patterns are especially good, with the front-to-back ratio exceeding 20 dB on all but one channel and showing very little of the many minor lobes and peculiar pattern distortions that arise in complex antennas.

The Jerrold VIP-series antennas (Fig. 2-24) use an unusual type of dual-mode director for all directors except the one nearest the

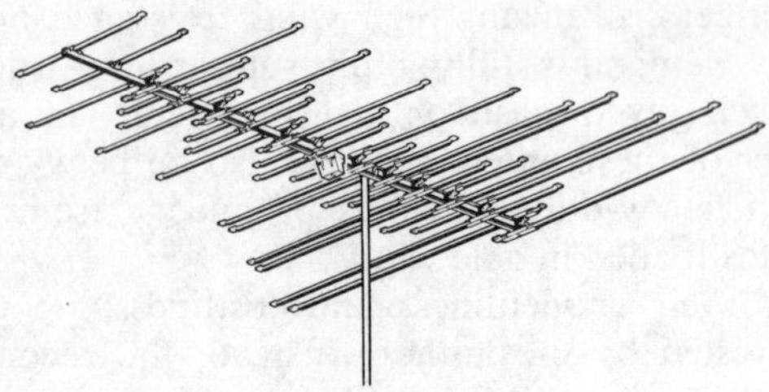

(*A*) *CH-4053.*

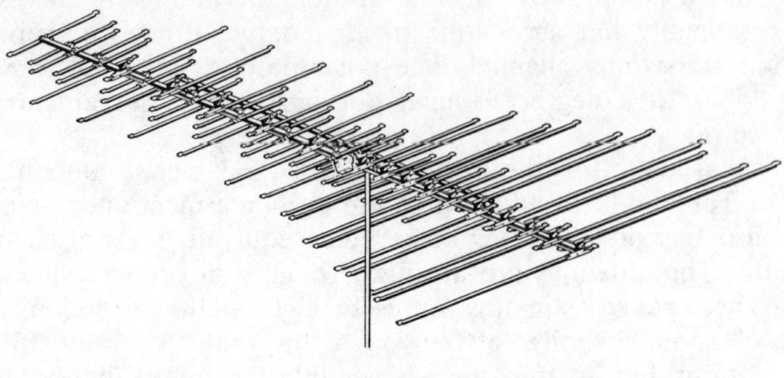

(*B*) *CH-5200.*

Courtesy Winegard Co.

Fig. 2-21. Winegard CH-4053 and CH-5200 antennas.

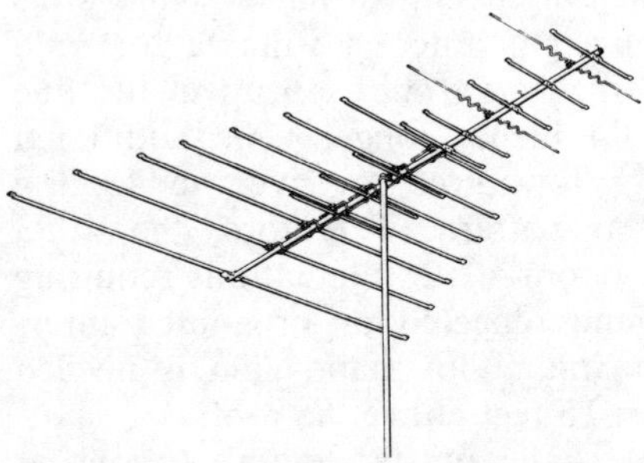

Fig. 2-22. Winegard X-220 antenna.

Courtesy Winegard Co.

driven elements. These unique director units each act as *two* collinear low-band directors and as *four* collinear high-band directors. This is why the VIP-307 has such a large director count. Another interesting feature of this type of dual-mode director is that the low-band resonant frequency is 108 MHz instead of 88 MHz as for the usual high/low director. This produces very high gain throughout the fm band—one reason why the VIP series has excellent fm performance.

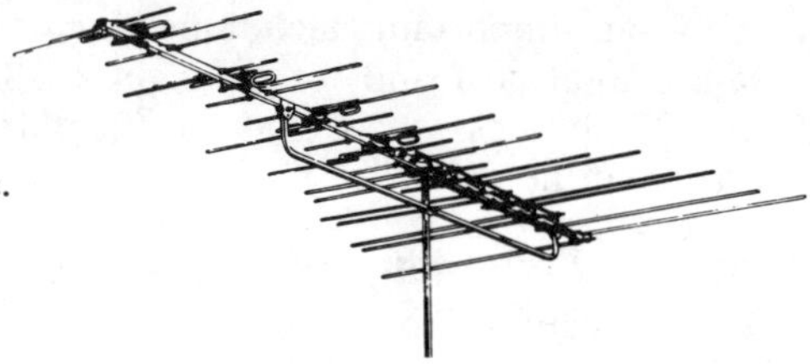

(*A*) *Antenna construction.*

Courtesy RCA

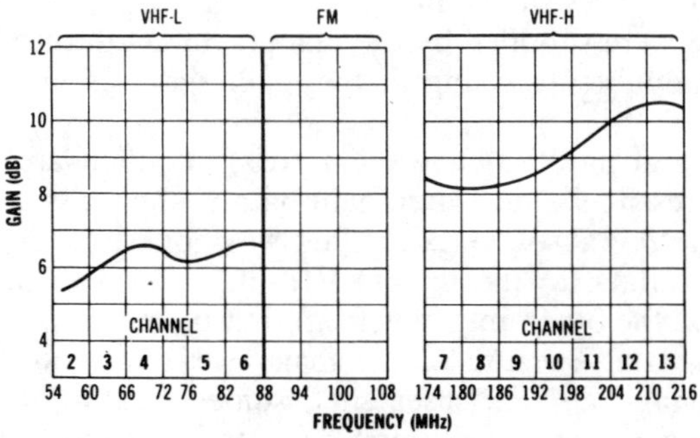

(*B*) *Gain curves.*

Fig. 2-23. RCA 3BG27 antenna.

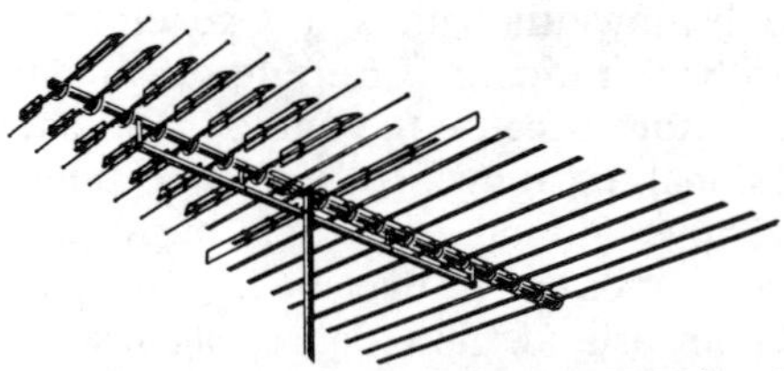

Courtesy TACO/Jerrold

Fig. 2-24. Jerrold VIP-307 antenna.

The operating principle of the "2–4" director is shown in Fig. 2-25. The "looped" middle section of each element reverses the phase of the antenna currents during high-band operation so the currents in both half-sections of each element are in phase. On the low band the entire element acts as a simple half-wave element.

At the end of the list of multiband vhf antennas are the single-band and single-channel antennas. These are included here to dramatize the increased performance possible from restricted bandwidth an-

Fig. 2-25. Jerrold "2–4" multimode director.

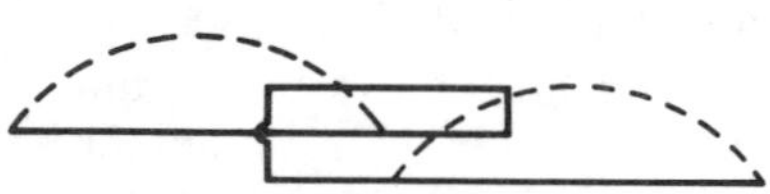

tennas—an important technique for weak-signal reception. Note that the total cost and total weight of the combined CH-2026 and CH-2073 are less than that of the CH-5200, although the gains average out the same.

Measured Performance Comparison

In this section, the vhf antennas measured by the author are compared as to physical characteristics and measured electrical performance. The prices shown in Table 2-2 are suggested list and dealer prices. These were supplied by the manufacturers during 1976. While they have undoubtedly changed, they still serve as an indication of relative cost.

The figures in the "Relative Gain (dB)" column are the author's attempt to present the measured gain data in the most useful manner by *approximating* absolute gain. This was done by using the manufacturer's gain data for the Jerrold VIP-303, and then calculating the gain figures of the other antennas from that by means of the measured relative gain data. Because of the many variables and difficulties of making antenna gain measurements, some of this data differs significantly from the manufacturers' data. It is therefore intended primarily to show how a variety of antennas compare when measured under *typical operating condtions* at the same time, same place, and with the same equipment.

The measured beamwidth figures are generally greater, and the measured front-to-back ratios generally less than those indicated on manufacturers' literature because the author's measurements were not made on a professional test range. If these same antennas are mounted on towers high above ground and roof, narrower beamwidth and higher front-to-back ratios can be expected.

Comparing the measured gain figures against the manfacturers' data reveals an interesting thing: the antennas whose measured low-band performance apparently fell short of the specifications have relatively short booms for the number of elements involved. A short-boom antenna will have lower output, lower front-to-back ratio, and wider beamwidth than a long-boom antenna of identical gain and pattern characteristics (based on professional test-range measurements where conditions approximate free space) when they are measured under "rooftop" conditions. This is because the long-boom antennas have much narrower vertical-plane patterns, so their performance is not affected by the presence of a roof or ground as much as a short-boom antenna is affected. The Winegard CH-4052 and Jerrold VIP-303 are good examples of this. The CH-4052 has *slightly* higher low-band gain than the VIP-303, but the measured results (made with the antenna mounted 12 feet above the roof) are generally lower, particularly on Channel 2. This is because the low-

Table 2-2. Comparison of Measured Antenna Performance

Antenna	Boom Length (in)	Weight (kg)	Prices ($)		Relative Gain (dB)							Beamwidth (°)				Front-to-Back Ratio (dB)			
			List	Dlr	2	4	5	7	9	11	13	2	5	7	13	2	5	7	13
Winegard CH-4210	65	1.5	32	19	−8.7	−6.0	−0.8	0.4	4.2	5.8	6.7	70	72	51	38	14	12	18	16
RCA 3BG09	56	1.5	22	13	3.6	−0.5	2.3	4.8	4.8	6.3	6.9	81	70	64	66	13	12	13	18
Antennacraft Mark-8	50	1.2	24	14½	1.0	3.7	3.0	7.3	8.1	7.4	6.9	90	72	35	43	6.4	12	7.5	17
RMS DJR-6	80	1.6	35	21	2.1	4.5	2.3	8.2	8.1	5.6	6.1	85	69	34	44	17	17	23	8.4
A.C.A. AC-511	71	1.9	35	21	3.2	3.4	3.2	7.1	6.8	6.8	7.2	90	76	35	53	12	18	22	15
Jerrold ZIP-12V	122	2.7	37	22	2.0	3.4	3.1	7.0	8.5	9.2	10.0	88	78	32	39	15	14	17	16
Winegard CH-4052	75	2.7	47	28	3.0	5.1	4.4	8.1	7.5	8.0	8.0	89	73	40	56	16	18	16	14
Jerrold VIP-303	85½	2.2	50	30	4.3	4.7	5.1	8.1	9.2	9.3	8.8	87	81	32	38	17	16	18	16

NOTE: Performance on the vhf low band is greatly influenced by the characteristics of the test site, because the distance between the antenna and its surroundings (roof, wires, structures) is small in terms of wavelength. In some cases the antenna output is enhanced on a particular channel (e.g., the RCA 3BG09 on Chan. 2), while in other cases the antenna output is decreased (e.g., Winegard CH-4210). If the same antennas are mounted at roof-to-antenna distances much greater than the 12 feet used for the author's measurements, the relative gain may be different.

band elements in the VIP-303 span 83 inches of boom length compared to only 61 inches in the CH-4052. The VIP-303 is less affected by the proximity of the roof and can develop the maximum output possible for the field strength. The unusually high Channel 2 output of the wide-spaced RCA 3BG09 (and 4BG-series antennas shown in Chapter 4) is due to this same factor.

CHAPTER **3**

UHF Antennas

The uhf tv band (Channels 14–83) is located between 470 and 890 MHz. Satisfactory tv reception is much more difficult to obtain at these frequencies than it is at vhf frequencies. There are several reasons for this: higher attenuation of uhf signals over the transmission path, lower sensitivity and poorer noise figure of uhf tuners, higher transmission-line losses, higher losses in signal-distribution components, and smaller antenna apertures. The line and component losses can be minimized by careful selection, and preamplifiers can solve the tuner problems. But antenna size is a tough problem. The smaller antenna size means that for a given field strength, a uhf antenna having a certain configuration will supply a much smaller signal voltage to the transmission line than will the equivalent vhf antenna. Another problem is that objects too small (in terms of wavelength) to cause signal reflections (ghosts) at vhf frequencies are large enough to do so at uhf frequencies. There is a bright side to the picture, however; the many different types of uhf antennas available offer a far greater choice of characteristics for combatting specific reception problems than are available from vhf antennas.

BOWTIE AND REFLECTOR ANTENNAS

The simplest antenna type having fairly consistent performance over the active portion of the uhf band is the bowtie with plane reflector. This antenna uses a full-wave driven element consisting of two triangular pieces of sheet aluminum arranged to form what looks like a large bowtie (Fig. 3-1A). The resulting fat dipole has very wide bandwidth, and a terminal impedance near 300 ohms. The driven element is spaced about a quarter wavelength in front of a plane reflector in the form of a steel grill. The use of a plane reflector instead of a rod reflector provides higher gain and excellent front-to-back ratio over a very wide bandwidth. The grill structure acts the same as a solid sheet because the grill rods are closely spaced in terms of wavelength at the highest frequency to be received. The open construction, however, greatly reduces the wind resistance of the reflector.

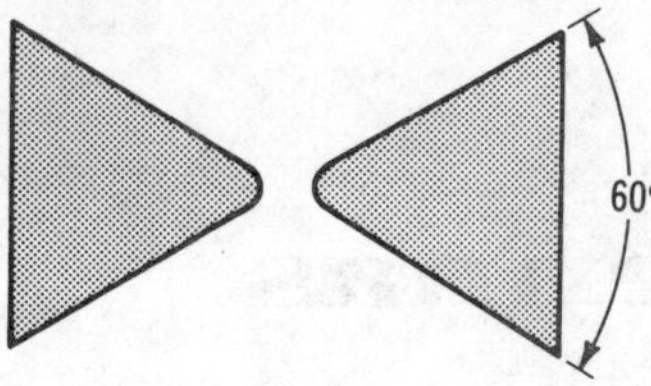

(*A*) *Sheet metal.*

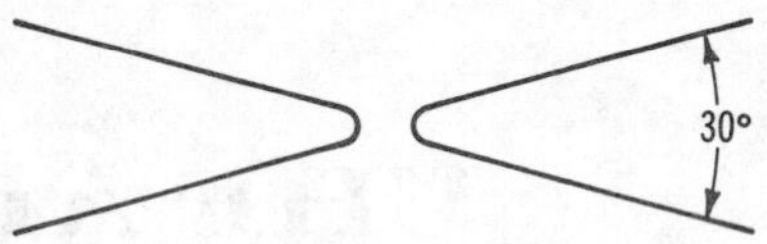

(*B*) *Wire outline.*

Fig. 3-1. Bowtie driven elements.

Single-Bay Bowtie

One bowtie element against a steel-grill reflector results in a small, lightweight, and low-cost antenna, capable of providing about 5–8 dB gain over the active uhf band. A representative example of the bowtie-reflector antenna is the JFD Electronics UHF-600, shown in Fig. 3-2A. Its polar pattern is comparatively broad at the lower end of the band, but gradually narrows with increasing frequency (Fig. 3-2B). This antenna is suitable for locations close to the transmitter where an unobtrusive and lightweight antenna is desired. In fact, the UHF-600 is ideal for use as an *indoor* antenna. As such it will perform rings around the uhf loop antenna normally supplied with tv sets.

4-Bay Bowtie

Four "bowtie" dipoles can be connected together and operated against a common plane reflector to produce a high-performance antenna popularly called a 4-bay. To lighten the antenna and reduce wind resistance, each bowtie element consists of a wire outline (Fig. 3-1B) instead of a sheet-aluminum triangle. The apex angle of the wire triangle is smaller to raise the impedance of each bowtie, so the combination of the four dipoles yields a good match at the antenna output terminals to 300-ohm transmission line.

The plane reflector of a 4-bay is usually about 4½ square feet of hardware cloth (large-mesh screen) or a steel grill. This results in a medium-sized antenna with excellent polar patterns (high front-

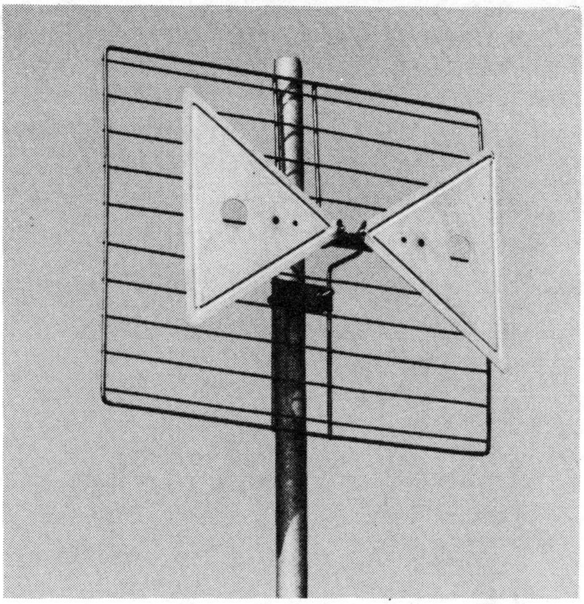

(*A*) Photograph of antenna.

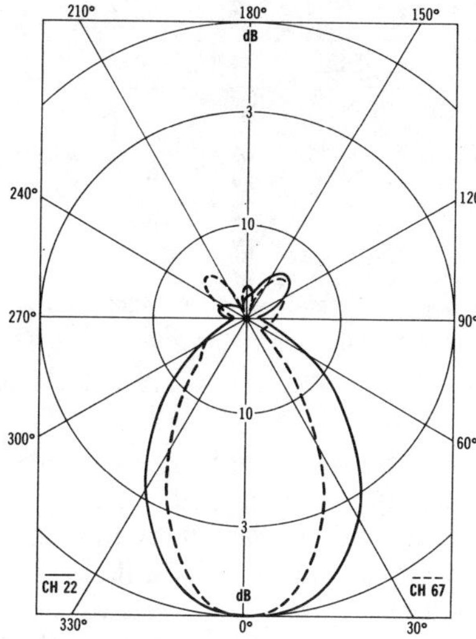

(*B*) Polar patterns.

Fig. 3-2. JFD UHF-600 antenna and polar patterns.

to-back ratio and no side lobes) and a high gain over the full uhf band.

A gain of 10–13 dB from Channels 14–83 is typical for a 4-bay, although the actual shape of the frequency-response curve will vary considerably from one manufacturer to another. The 4-bay is sold by more manufacturers than is any other basic design. Nearly any model will yield acceptable performance; the ones shown here provide excellent performance.

The Antenna Corp. of America's AC-320 (Fig. 3-3A) is the largest of the 4-bays; its reflector screen measures 36×20 inches. This screen is also the most opaque to radio waves, since the mesh spacing is only 1 inch in the vertical direction. These factors mean the screen most closely approximates the theoretical infinite conducting-sheet reflector. An interesting feature of this antenna is that

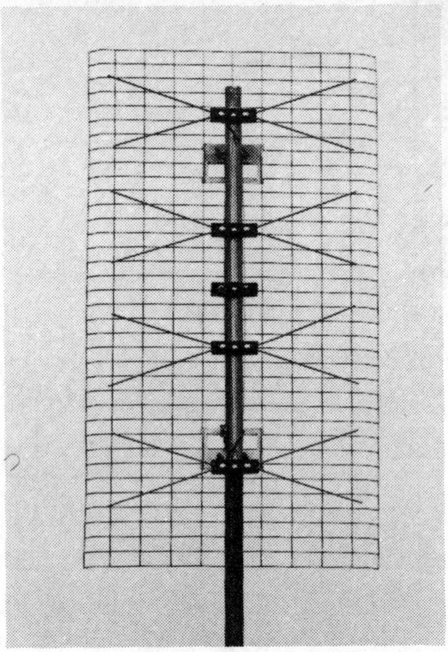

Courtesy Sony Corp. of America

(*A*) *Photograph of antenna.*

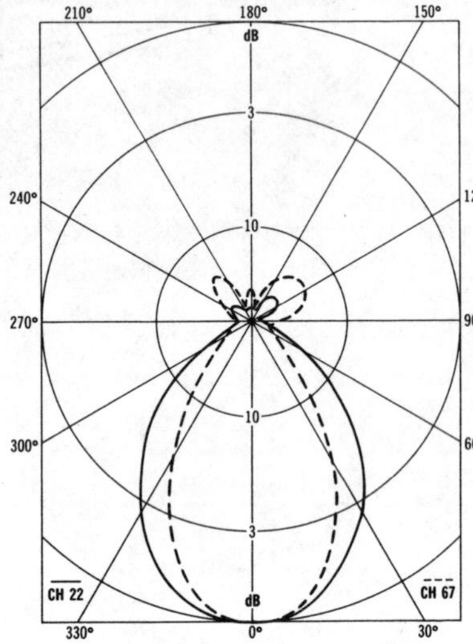

(*B*) *Polar patterns.*

Fig. 3-3. A.C.A. AC-320 antenna and polar patterns.

it can be converted into a bidirectional antenna (figure-8 polar pattern) by removing the reflector screen. (Pry open the tabs securing the mounting bracket to the screen.) If ghosts are no problem and the signal strength is sufficient, this allows rotatorless reception of stations in opposite directions.

Winegard's KU-420 is identical with A.C.A.'s AC-320. Winegard also offers a special model for optimum reception of translator channels. This antenna, the KT-420, has shorter element lengths and spacing to peak the gain over Channels 50–83.

The Lance KW4S (Fig. 3-4A) has a 34 × 19-inch steel reflector screen made of horizontal rods welded to a heavy-gauge frame. This

SAMS

Book Mark

Sams books cover a wide range of technical topics. We are always seeking recommendations from our readers regarding their informational needs. Please complete this questionnaire and return it to us with your suggestions. We appreciate your comments.

1.) Do you own/use a computer?

- ☐ No.
- ☐ Yes. Please specify brand and model:
- ☐ Apple ______
- ☐ Atari ______
- ☐ Coleco ______
- ☐ Compaq ______
- ☐ Commodore ______
- ☐ Hewlett-Packard ______
- ☐ IBM ______
- ☐ MSX ______
- ☐ Panasonic ______
- ☐ Radio Shack ______
- ☐ Sanyo ______
- ☐ Texas Instruments ______
- ☐ Timex Sinclair ______
- ☐ Other (specify) ______

2.) Do you own/use any software, systems, or languages?

- ☐ No.
- ☐ Yes. Please specify brand/type:
- ☐ Word Processing ______
- ☐ Spreadsheet ______
- ☐ Graphics/Sounds ______
- ☐ Games/Entertainment ______
- ☐ Data Base Management ______
- ☐ Integrated Software ______
- ☐ Operating Systems (specify) ______
- ☐ Computer Languages (specify) ______
- ☐ Other (specify) ______

3.) Are you interested in any of the following electronics or other technical topics?

- ☐ No.
- ☐ Yes. Please specify:
- ☐ Acoustics, Speech, Signal Processing
- ☐ Aerospace and Electronic Systems
- ☐ Alternative Energy Research
- ☐ Antennas and Propagation
- ☐ Communications
- ☐ Consumer Electronics
- ☐ Control Systems
- ☐ EMC/EMI
- ☐ Information Theory
- ☐ Instrumentation and Measurements
- ☐ Manufacturing Technology
- ☐ Power Engineering
- ☐ Robotics and Artificial Intelligence
- ☐ Security Electronics
- ☐ Service Guides—Electronics
- ☐ Solid State Circuits, Components, Hybrids
- ☐ Other ______

4.) Occupation

- ☐ Bus. Professional (specify) ______
- ☐ Engineer (specify) ______
- ☐ Technician (specify) ______
- ☐ Scientist
- ☐ Educator
- ☐ Military
- ☐ Health Professional
- ☐ Programmer/Analyst
- ☐ Student
- ☐ Hobbyist
- ☐ Other (specify) ______

5.) Education

- ☐ High School
- ☐ Technical School
- ☐ College Graduate
- ☐ Post Graduate

Comments ______

Name ______

Street Address ______

City ______

State ______ Zip ______

Daytime Telephone (______) ______

SAMS

21584

SAMS™ Book Mark

Book Mark SAMS™

NO POSTAGE
NECESSARY
IF MAILED
IN THE
UNITED STATES

BUSINESS REPLY CARD

FIRST CLASS PERMIT NO. 1076 INDIANAPOLIS, IND.

POSTAGE WILL BE PAID BY ADDRESSEE

HOWARD W. SAMS & CO., INC.
ATTN: Public Relations Department
P.O. BOX 7092
Indianapolis, IN 46206

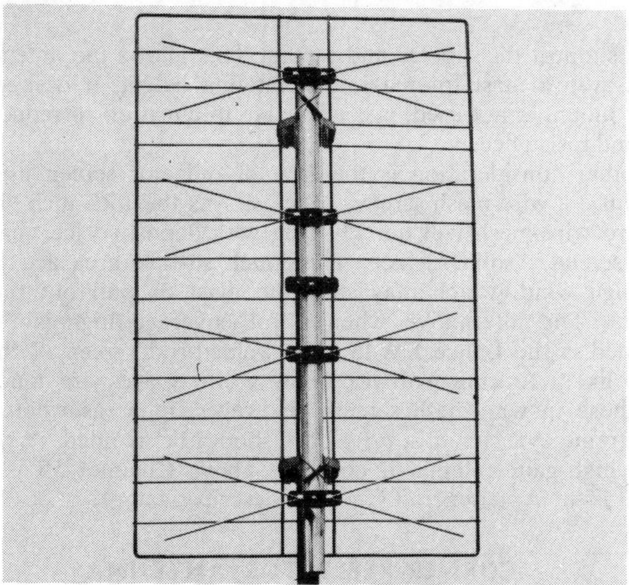

Courtesy Sony Corp. of America

(*A*) *Photograph of antenna.*

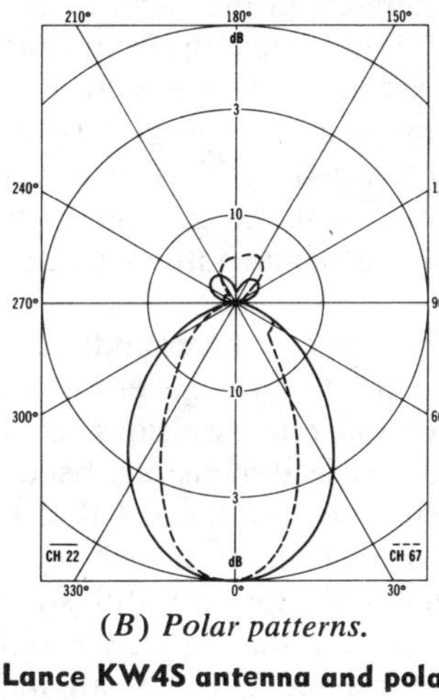

(*B*) *Polar patterns.*

Fig. 3-4. Lance KW4S antenna and polar patterns.

antenna has a rising gain characteristic that peaks around Channel 65, although gain remains very high throughout the translator channels. Like the A.C.A./Winegard antennas, the Lance has very clean polar patterns and a high front-to-back ratio (Fig. 3-4B). This same antenna is available from Jerrold Electronics as their Model 3044.

Various mechanical aspects of the 4-bays may affect your choice. The Lance KW4S uses a 1¼-inch-diameter steel tube with swedged end to support the dipoles and screen. This allows the antenna to be jam-fit atop a mast for maximum antenna height, if desired. If the jam-fit mount is not used, it is advisable to cut it off to reduce weight and wind resistance.

Another consideration is the type of reflector screen used. Most 4-bays use a wire-mesh screen. In ice storms the 1×2-inch "holes" in standard wire-mesh will quickly clog with deposited ice, making the entire screen a solid surface. This much surface area acts as a sail in a high wind, which may snap the mast or pull out the mount anchors. The alternative when a full-coverage high-gain antenna is needed is the Lance KW4S. The welded-rod screen of this 4-bay is less likely to clog with ice because the spaces are much larger than those in wire-mesh screens. However, in areas where icing is too extreme even for this, *all* 4-bays should be avoided. A good substitute high-gain antenna if coverage above Channel 55 is not important is the Antennacraft G-1483 (described later).

CORNER-REFLECTOR ANTENNA

The corner-reflector antenna is a very simple antenna consisting of a full-wave driven element in the form of a sheet-aluminum bow-tie, spaced about half a wavelength from the apex of a reflector formed by two planes set at right angles to each other. The angle of the reflector produces medium-to-high gain by narrowing the *vertical-plane* beamwidth of the antenna. This also makes the antenna less susceptible to reflections from the ground, nearby objects, and aircraft. The horizontal-plane beamwidth is relatively unaffected by the reflector angle.

The dimensions of the reflector profoundly affect the antenna's gain vs. frequency characteristic. Gain is generally increased by increasing the length and width of the reflector sides, although the rate of improvement tapers off as these dimensions become long compared to wavelength. In practical cases this means that gain is very much dependent on reflector size at the low end of the band, where the dimensions of the usual reflector are still small in terms of wavelength. At the high end of the uhf band, the reflector dimensions are big enough in terms of wavelength that an increase in size doesn't matter that much. This situation will be evident when the "large" and "small" RCA antennas are compared later on.

Most corner-reflector antennas now use self-supporting aluminum-rod construction for their reflectors. The spacing between the reflector rods has greatest effect on the front-to-back ratio at the high end of the uhf band. A spacing of 2 inches or less is needed to maintain a high front-to-back ratio at Channel 83, while a spacing as great as 4 inches will suffice at Channel 14.

Parabolic-Cylinder Reflector

A recent variation of the corner-reflector antenna uses a reflector approximating a parabolic cylinder. No significant gain improvement results from doing this with normal reflector sizes. High gain is obtainable from a parabolic reflector only if it functions in an optical mode. For this to occur the distance from driven element to reflector must be much greater than a wavelength, and the reflector must be many wavelengths high. This requires a reflector height of over 5 feet at uhf tv frequencies.

Commercial Antennas

Many manufacturers sell corner-reflector antennas, but the outstanding model is the RCA 7B140 (Fig. 3-5A). This low-cost antenna had the best performance of the many corner-reflector antennas tested, superb construction, and an attractive appearance too! It also has an interesting feature found on few of the other models: a central reflector rod. This fills in the "hole" in the reflector screen that is characteristic of most corner reflectors using rod screens, and thereby improves the front-to-back ratio at high frequencies. Another interesting point is that the corner-reflector angle is around 100°, instead of the usual 90°.

The polar patterns (Fig. 3-5B) of the 7B140 progress smoothly from a moderately broad lobe at the low end of the band to a fairly narrow one around Channel 60. Then the pattern broadens again. Correspondingly, the gain rises with frequency to a peak around Channel 60 (see the solid curve in Fig. 3-6).

The 7B140 belongs to the "small" corner-reflector category, like the other models it was compared against. The RCA 7B141 is its big brother, having a frontal area of 1060 square inches compared to the 660 square inches for the 7B140. The only difference physically is the reflector size; the dipole dimensions and spacing are identical. The larger reflector size produces 1–2 dB higher gain over Channels 14–45. The broken gain curve in Fig. 3-6 shows how the 7B141 compares to the 7B140 (solid curve). Obviously the bigger antenna is useful only where the channels of interest (particularly the weak ones) are below Channel 45; it is pointless to employ the larger corner reflector if the weak channels in your area are in the upper portion of the uhf band. This frequency response curve is typical to large corner reflectors; the manufacturer's gain curves for the Channel Master 4193 are similar to those for the 7B141.

Another good corner reflector with useful and unusual characteristics is the Gavin CR-5 (Fig. 3-7A). This antenna provides 7–10 dB gain over the *entire* uhf band (Channels 14–83), a respectable performance considering its rather small size. Unlike most other an-

Courtesy Sony Corp. of America

(A) Photograph of 7B140.

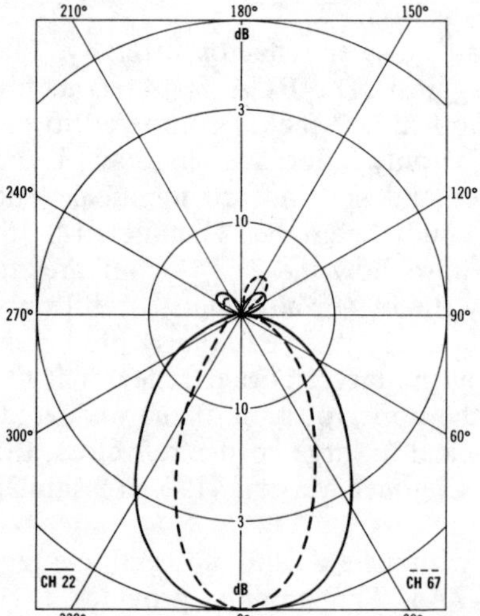

(B) Polar patterns of 7B140.

Fig. 3-5. RCA 7B140 antenna and polar patterns.

tennas, the gain of the CR-5 does not fall off after Channel 70, but remains near the 10-dB mark. The beamwidth is very consistent with frequency (Fig. 3-7B), something unusual for corner reflectors.

The CR-5 does not lock into its deployed position like other corner reflectors; the reflector arms are held open by pressure from the antenna mast. This allows you to refold it immediately after removing it from the mast. This feature and its small size make it ideal for use by antenna installers as a test antenna to determine uhf signal strength at new locations. After making the survey and finding the best place

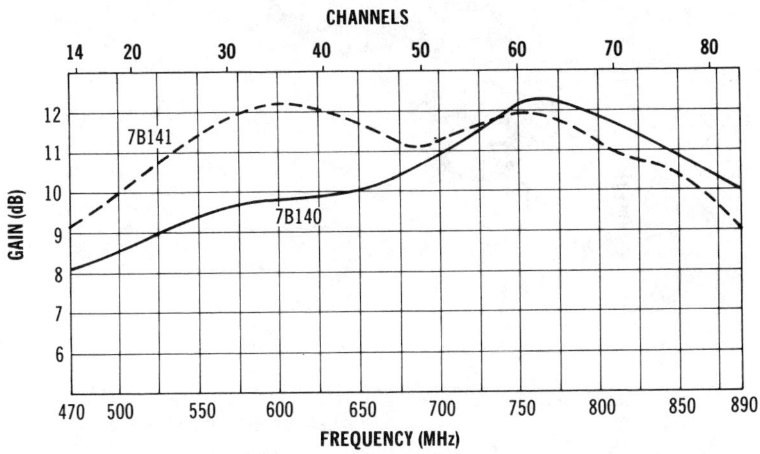

Fig. 3-6. Gain comparison of large and small corner reflectors.

to mount the permanent antenna, the installer can fold the CR-5 and stow it in his truck. This antenna is the lowest priced of the corner reflectors, whether purchased from a Gavin distributor or from Radio Shack as Model 15-1629.

CORNER-REFLECTOR YAGI

The corner-reflector yagi is a hybrid antenna type, the result of combining the corner reflector with the many directors of the yagi, in an attempt to produce a high-gain broadband antenna. Unfortunately, few of the many models offered succeed despite exotic appearances and high prices. The problem is that most directors boost gain greatly and narrow the beamwidth significantly over a relatively small frequency range. At progressively lower frequencies director-produced gain gradually decreases. At frequencies higher than the director design frequency the gain decreases very rapidly; in fact, at much higher frequencies the directors actually reduce gain. Be-

Courtesy Sony Corp. of America

(A) Photograph of antenna.

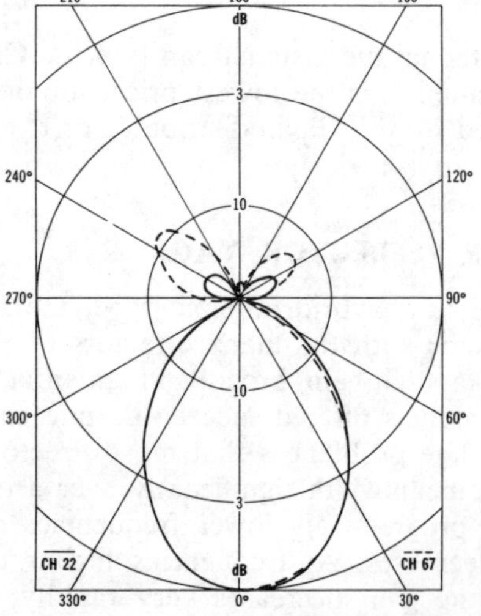

(B) Polar patterns.

Fig. 3-7. Gavin CR-5 antenna and polar patterns.

Courtesy Sony Corp. of America

(A) Photograph of antenna.

(B) Polar patterns.

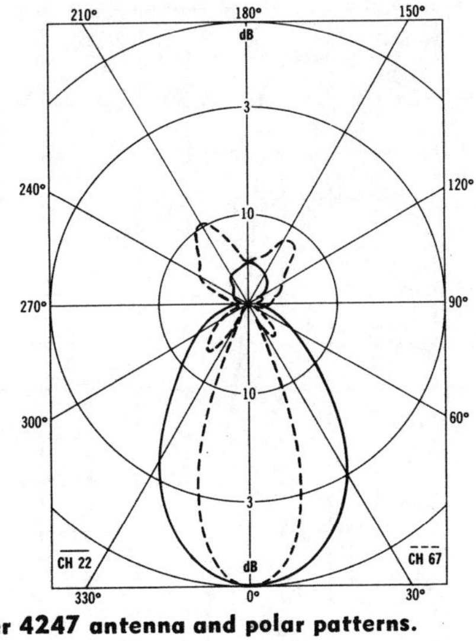

Fig. 3-8. Channel Master 4247 antenna and polar patterns.

cause of this, directors on a full-coverage (Channels 14–70) antenna are usually peaked around Channel 60, and have little effect at the low end of the uhf band. Therefore, the low-end performance of the usual corner reflector yagi is essentially that of a plain corner reflector, while the high-end performance falls off sharply in the Channel 70–83 region.

An excellent full-coverage corner-reflector yagi is the Channel Master 4247 (Fig. 3-8A). The fine performance of this antenna is due mainly to its diamond-shaped directors. Their high width-to-length ratio makes these directors far more broadband than the rods used on most other models, so the gain-increasing effect of the directors extends over a much wider frequency range. The directors are arranged in collinear pairs, so all 14 directors and the corner reflector are accommodated by a relatively short boom (47 inches). The net result is an antenna with a claimed gain of 8–10 dB over Channels 14–70. However, the author's own measurements (reported at the end of this chapter) indicate much higher gain than this.

The beautifully constructed 4247 has vhf feedthrough provisions, allowing the use of one transmission line for separate vhf and uhf antennas. To use this feature, a short length of twinlead is connected between the terminals of the vhf antenna and the upper terminals of the driven element of the 4247 (a three-quarter–wavelength folded dipole). Then the transmission line to the tv set or signal-distribution

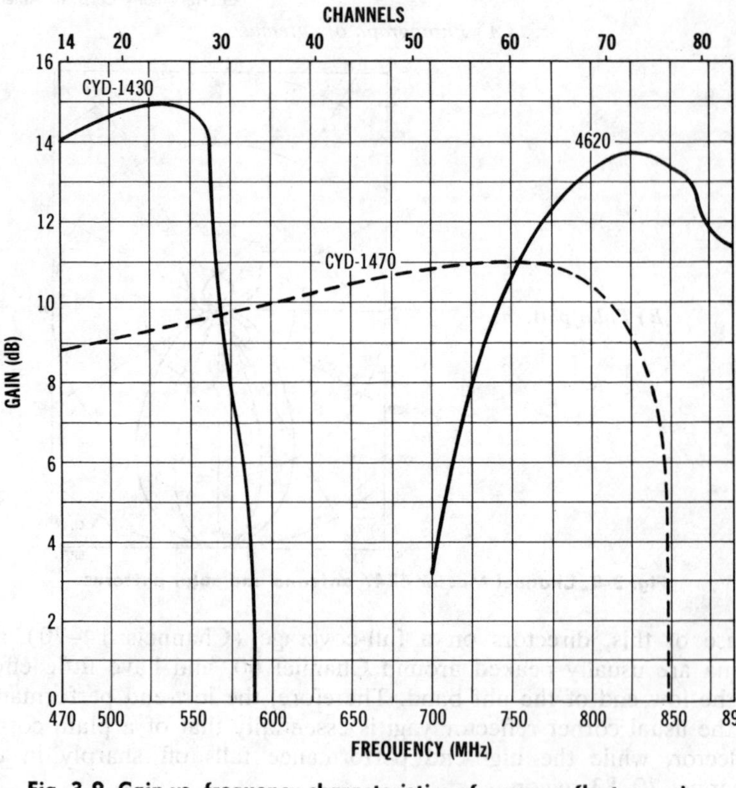

Fig. 3-9. Gain vs. frequency characteristics of corner-reflector yagis.

system is connected to the lower terminals of the driven element of the 4247.

The polar patterns of the Channel Master 4247 (Fig. 3-8B) are typical of the corner-reflector yagi. The beamwidth is fairly broad at the low end of the band, but becomes very narrow at the high end of the band where the directors are most effective. As a consequence of the 14 directors, the 4247 has very narrow high-end beamwidth (about 32°). The side lobes in the Channel 67 pattern are harbingers of pattern breakup at higher frequencies, also a consequence of the 14 directors. This antenna, like most corner-reflector yagis not specifically designed for that service, should not be used in the translator range (Channels 70–83). The Channel Master 4260 (next section) would be a much better choice for this application.

Narrow-Band Antennas

The most successful implementation of the corner-reflector yagi is for limited frequency-range antennas. Over a narrow frequency range (15–20 channels) the effects of a long boomful of directors are spec-

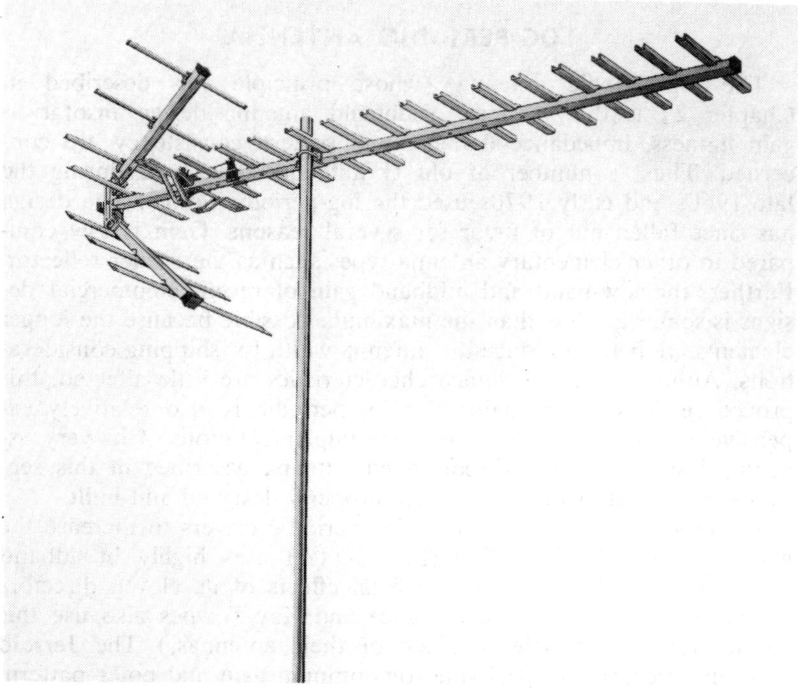

Courtesy TACO/Jerrold

Fig. 3-10. Structure of Jerrold CYD-1430 and CYD-1470 antennas.

tacular. High gain and narrow beamwidth result from the many directors peaked for the frequency range of interest. High gain and high front-to-back ratio result from the corner reflector. The combination of the two yields extremely high gain for a given antenna weight when the reflector and driven-element dimensions are also tailored to the frequency range of interest.

Fig. 3-9 shows how frequency coverage can be traded for gain. The solid surve at the left is for the Jerrold CYD-1430, an antenna covering only Channels 14–27. The dashed curve in the center is for the CYD-1470, a full-coverage antenna. Both antennas are constructed the same: 14 directors on a 79-inch boom (Fig. 3-10). The sole difference is in the element dimensions. Yet while the 1470 provides only 9–11 dB gain, the 1430 provides a spectacular 14–15 dB gain over Channels 14–27, with an antenna weight of less than a kilogram!

Limited-frequency coverage antennas are often used for translator service (Channels 70–83). Channel Master's Model 4260 is one such antenna. As shown by the solid curve at the right in Fig. 3-9, it provides 11½–13½ dB gain over Channels 62–83.

LOG-PERIODIC ANTENNA

The log-periodic antenna (whose principle was described in Chapter 2) is the "perfect" wideband antenna design insofar as gain flatness, impedance flatness, and pattern consistency are concerned. Thus, a number of uhf tv antennas developed during the late 1960s and early 1970s used the log-periodic design. The design has since fallen out of favor for several reasons. Gain is low compared to other elementary antenna types such as the corner reflector. Further, the low-band and midband gain of recent commercial designs is somewhat less than the maximum possible because the longer elements are bent to reduce the antenna width for shipping considerations. Although the impedance characteristics are little affected, this procedure does reduce gain. The log-periodic is also relatively expensive and difficult to build in some implementations of its very exacting design. The now-discontinued antenna described in this section was the last of those that were properly designed and built.

Directors were often added to log-periodic drivers to increase the gain. The Jerrold PAU-700 (Fig. 3-11A) uses highly broadband "butterfly" directors, so the beneficial effects of its eleven directors extend well down the band. (Lance and Kay-Townes also use this variation of the bowtie in some of their antennas.) The Jerrold butterfly directors are peakable for optimum gain and polar pattern. The polar patterns shown in Fig. 3-11B were made with the first eight directors shortened. Even so, the minor forward lobes (which

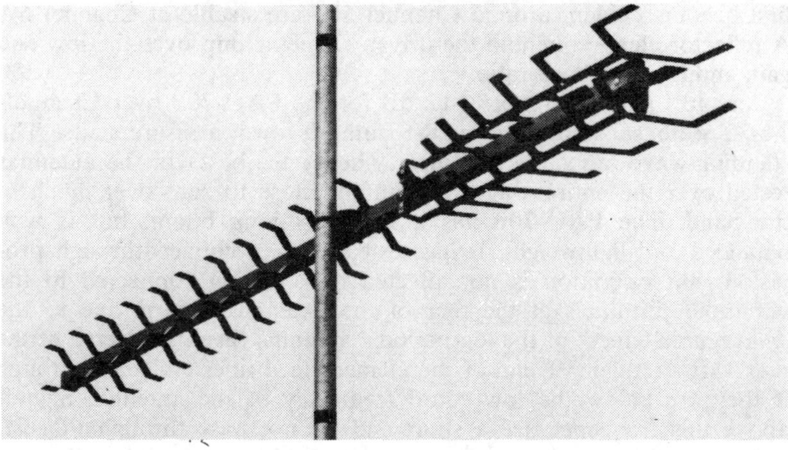

Courtesy Sony Corp. of America

(A) Photograph of antenna.

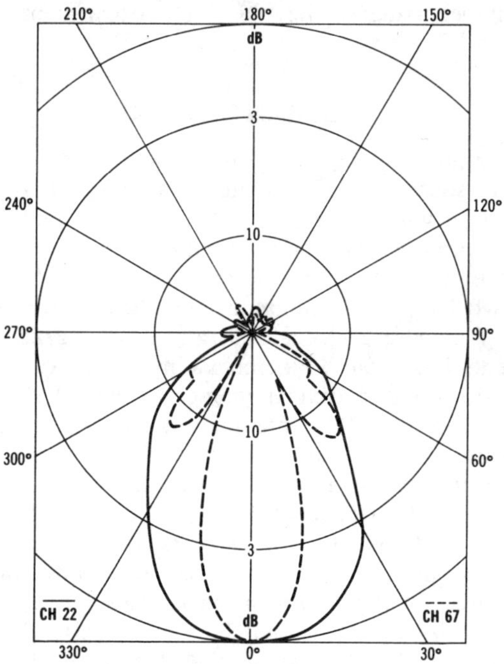

(B) Polar patterns.

Fig. 3-11. Jerrold PAU-700 antenna and polar patterns.

first become evident around Channel 50) are sizable at Channel 67. A reflector element behind the driven elements improves the low-end gain and front-to-back ratio.

Jerrold's gain claims of 8–12 dB for the PAU-700 over Channels 14-70 were substantiated by the author's own measurements. The standing-wave ratio was the lowest (hence the best) of the antennas tested over the entire uhf band, running close to zero over much of the band. The PAU-700 has a relatively long boom, but is well balanced and lightweight. It has a near-perfect vhf feedthrough provision; uhf operation is not affected by anything connected to the vhf input terminals at the rear of the antenna. This is due to the "waveguide effect" of the log-periodic antenna. Signals injected at the rear (low-frequency) end of the element feed lines will pass through if they are below the low cutoff frequency of the antenna. Signals above this frequency see a short and do not pass through the element feed lines to the output terminals at the front of the antenna. Thus, vhf signals can pass right through the uhf antenna to the common downlead, while uhf signals picked up by the vhf antenna are blocked. This prevents signal cancellation, ghosts, and pattern distortions from occurring to the uhf signal because of undesired uhf pickup on the vhf antenna.

HOVERMAN ANTENNA

The Hoverman antenna is the only new antenna design since the log-periodic to succeed as a consumer tv antenna. This unique design is often confused with the 4-bay, but its performance characteristics and operating principle are radically different.

The heart of the Hoverman antenna is the driven element. This consists of two lengths of aluminum wire bent into segments of approximately 7 inches each, and arranged in zigzag configuration. These segments are about 17 percent longer than one-quarter wavelength at the lowest operating frequency (Channel 14). In its commercial implementation, the Antennacraft G-1483 (Fig. 3-12A), the driven element is backed up by eight half-wave reflector elements arranged in four collinear pairs. Their lengths are chosen to provide maximum gain at the frequencies where the gain of the driven elements begins to fall off. The outer pairs are resonant at the lower edge of the uhf band; the inner pairs are resonant around midband. The overall effect of this combination is unusually high gain for an antenna of this size over Channels 25–55, and useful gain extending from Channels 14–60. In fact, the Antennacraft G-1483 has the highest gain of any antenna tested on Channels 25–55, yet it is the lightest of all! Although the front-to-back ratio becomes increasingly poor with increasing frequency (Fig. 3-12B) the front-to-side ratio

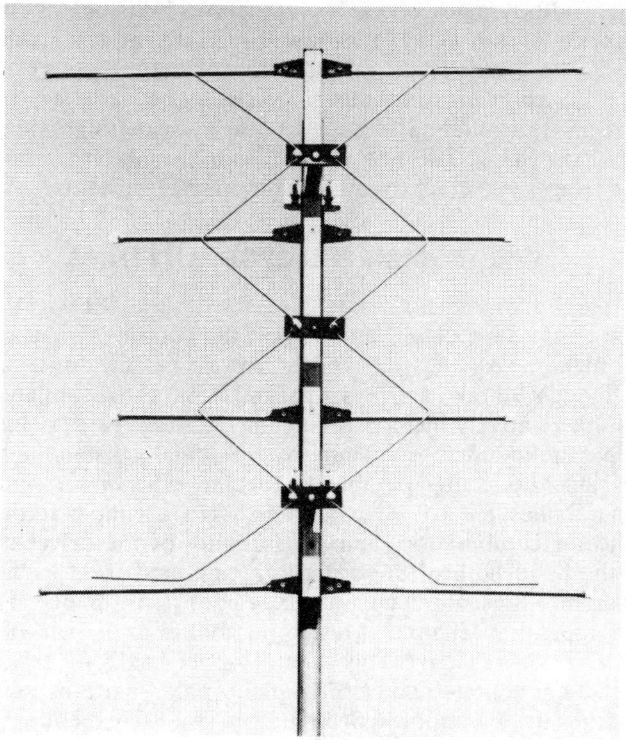

Courtesy Sony Corp. of America

(A) Photograph of antenna.

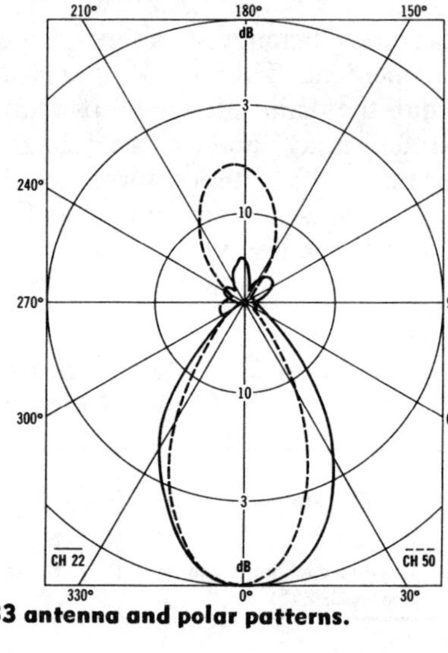

(B) Polar patterns.

Fig. 3-12. Antennacraft G-1483 antenna and polar patterns.

and beamwidth remain good at all frequencies in its operational range.

The price of the G-1483 is low, especially at retail where it is sold by Radio Shack as their Model 15-1627. If reflections from the rear are no problem, and the channels to be received are below Channel 55, this antenna is ideal for use in near-fringe areas. Moreover, its light weight and low wind resistance make it especially suitable for fringe-area stacked arrays (described in Chapter 8).

PARABOLIC-REFLECTOR ANTENNA

The parabolic-reflector antenna is the largest, heaviest, and the most expensive type of uhf antenna on the consumer market. It also has the highest absolute gain and narrowest beamwidth of any basic type, although billboard arrays (stacked 4-bays) of similar area will surpass this relatively inefficient antenna type in regard to gain.

The parabolic-reflector antenna type is actually a small antenna fed by the signal-collecting parabolic reflector. The *driver* antenna for small parabolics (3–4-ft diameter) is often a simple folded dipole and reflector combination. Since the quality of the driver so greatly affects the bandwidth of the overall antenna and front-to-back ratio, small parabolics usually have especially poor performance. The driver used in the large Channel Master parabolics is a pair of bowties against a screen reflector. This type of driver has good front-to-back ratio, and horizontal- and vertical-plane polar patterns suitable for illuminating the parabolic reflector. This results in better polar patterns for the overall antenna, and higher efficiency than is obtainable from the typical small parabolic.

The best parabolic antennas for consumer tv applications have an aperture efficiency of about 50 percent, while a 4-bay runs close to 100 percent. This means that a 4-bay (4½ sq ft surface area) will equal the gain of a 3½-ft-diameter parabolic antenna (9.4 sq ft surface area), and do the job at a much lower price. Furthermore, because of the low-performance drivers used in small parabolics, the overall antenna performance (front-to-back ratio, bandwidth, etc.) is correspondingly poor. What it all boils down to is this: avoid small parabolics like the plague. In the large sizes the parabolic

Table 3-1. Manufacturer's Performance Data for Channel Master 4251 Antenna

Characteristic	Frequency (MHz)					
	470	500	600	700	800	890
Gain (dB)	16	17.8	17.4	17.9	19.9	15.6
Front-to-Back Ratio (dB)	16	17.1	14.5	15.8	12	12.6
Beamwidth (°)	20	19	17	16	16	13

antenna becomes a feasible alternative to a billboard array. For example, the Channel Master Model 4251 7-foot-diameter parabolic has a surface area of 38 square feet, so it takes an array of four 4-bays to equal its gain. Although four 4-bays, a signal combiner, and phasing cables are slightly cheaper than the 4251, using the big parabolic eliminates a great amount of work building and phasing the billboard array.

The Channel Master 4251 and 4250 shown in Fig. 3-13 were not included in the author's comparative gain tests. However, Channel Master lab data is summarized in Table 3-1.

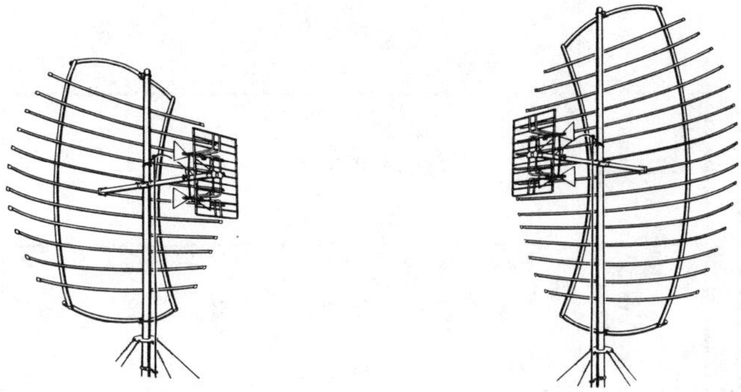

(*A*) *Model 4250, 6 feet, $73 list, $44 dealer.*

(*B*) *Model 4251, 7 feet, $89 list, $53 dealer.*

Courtesy Channel Master

Fig. 3-13. Channel Master parabolic antennas.

ANTENNA PERFORMANCE COMPARISON

In the preceding sections of this chapter, the best of each antenna type were described. In this section, the antennas are compared as to physical characteristics and electrical performance, and in regard to suitability for specific applications.

Table 3-2 contains physical and financial data not previously given, and a tabulation of the −3-dB beamwidths and front-to-back ratios (taken from the polar patterns). The prices shown in this table are suggested list (resale) prices and either actual or suggested dealer prices. These prices were supplied by the manufacturers during May 1976. While they have undoubtedly changed, they still serve as an indication of relative price.

Fig. 3-14 is the author's attempt to present the measured gain data in the most useful manner by *approximating* absolute gain curves and frequency response. The curves were determined by using the manufacturer's gain data for the Gavin CR-5, and then plotting the other

Table 3-2. Comparison of Characteristics

Antenna	Physical				Approx. Price ($)		VHF Feed-through	Measured					
								Bandwidth (°)			Front-to-Back Ratio (dB)		
	Wt (kg)	L (in)	W (in)	H (in)	Dlr	Ret		22	50	67	22	50	67
1. Bowtie & Reflector													
JFD UHF-600	0.6	4	18	15	5	8½	No	60		43	15		15
2. Log-Periodic													
Jerrold PAU-700	0.85	58½	7	4	14	23	Yes	58		26	22		20
3. Corner-Reflector													
CR-5	0.75	14	18	26½	5¾	10	No	59		57	18		9
RCA 7B140	0.85	14	22	30	6½	11	No	63		39	20		17
RCA 7B141	1.1	16½	28	38	8½	14	No	53		48	19		20
4. Corner-Refl. Yagi													
Ch. Master 4247	1.1	47	19	23	15	22	Yes	55		32	18		10
5. 4-Bay Bowtie													
Lance KW4S	1.2	5½	19¼	34	9	16	No	64		41	18		15
A.C.A. AC-320	1.2	5½	19¼	36	10	17	No	64		45	19		14
6. Hoverman													
Antennacraft G-1483	0.6	5	30	30	7½	13	No	50	40		16	6	

Table 3-3. Antenna Selector Guide

Characteristic	UHF-600	AC-320	KW4S	CR-5	7B140	7B141	4247	PAU-700	G-1483
High Gain, 14–70		●	●				O		
High Gain, 14–50		O	O			O			●
High Gain, 50–83		O	●		O				
High FBR & FSR, 14–70					O	O		●	
High FBR & FSR, High-End					O			●	
High FBR & FSR, Low-End		O	O	O	O	O	O	●	
Narrow Beamwidth, 14–70							O	●	
Narrow Beamwidth, High-End							O	●	
Narrow Beamwidth, Low-End						O	O		●
Wide Beamwidth, 14–17				●					
Wide Beamwidth, High-End				●					
Wide Beamwidth, Low-End		●	●		●				
Low Cost	●			●	O				O
Small Size	●								
Light Weight	●			O					●
VHF Feedthrough							O	●	
Low Wind Resistance									●
Low Ice Loading				O					●

● Highest rating O High rating

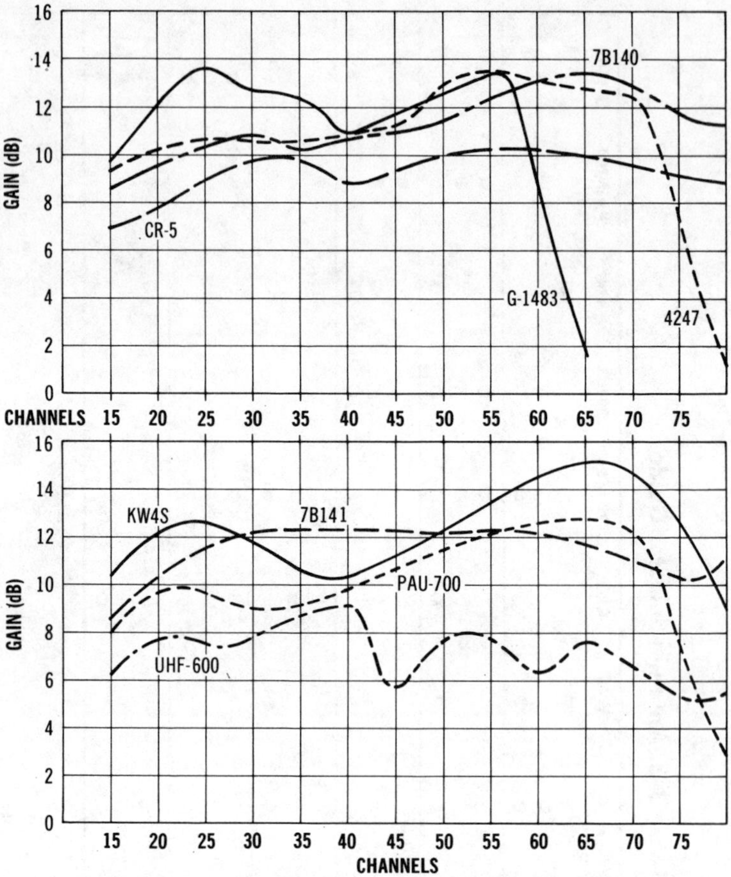

Fig. 3-14. Gain vs. frequency curves for various antennas.

antenna curves from that by means of the measured relative gain values. Because of the many variables and difficulties of making antenna gain measurements, some of these curves differ somewhat from the manufacturers' data. However, this data is far superior to manufacturers' data for comparing different antennas, because this gain data was compiled on the same test range, on the same day, with the same equipment, and by the same person.

Table 3-3 facilitates picking the antenna whose characteristics best suit the requirements of your application. Keep in mind, however, that this table is just a rough guide; additional factors may be pertinent in choosing the best antenna for your application.

For some applications, an ordinarily desirable characteristic may be undesirable. For instance, a high-gain antenna within 10 miles of the transmitter is most likely a waste of money if it is used to feed only one tv set. In fact, the very high signal input to a uhf tuner that occurs under these circumstances may cause cross modulation, since there is no protective gain-control element between the antenna terminals of the tuner and mixer diode. A medium-gain antenna with excellent patterns, like the RCA 7B140, is a better choice in this case.

Similarly, narrow beamwidth will minimize the pickup of ghosts or interference arriving from a direction close to that of the desired signal. But an antenna rotator is required if there is more than one station to be received and the stations lie along different bearings. For a situation like that a wide-beamwidth antenna will allow a rotatorless installation, and thereby save a lot of money. A corner reflector or 4-bay is suitable, depending on the gain required.

CHAPTER 4

All-Channel Antennas

The all-channel (Channels 2–83) antenna is the most popular category of antenna sold in the U.S. today. This is because all-channel antennas are convenient to erect and install, weigh less than separate antennas, require less mast space, and are less noticeable than separate uhf and vhf antennas. The "hitch" in this pleasant situation is that for any given antenna size, the performance on any band is naturally inferior to a separate vhf antenna of the same boom length and price, and the uhf section is often a low-performance "afterthought."

Selecting the proper vhf-uhf antenna for your situation is a lot harder than selecting individual antennas for two reasons. First, most all-channel designs require compromises in the design of one or both sections. Second, the requirements for each section may be greatly different in regard to gain and pattern. In difficult reception situations it therefore becomes highly unlikely that you can find one antenna where each section meets very exacting performance requirements. Fortunately, most reception situations in metropolitan and suburban areas are not that critical, and a suitable all-channel antenna for such locations can be found. In weak-signal areas the situation is different. If the signal is so weak that a high-performance antenna is needed, separate vhf and uhf antennas are the *best technical approach and the most practical.* This way boom lengths are manageable and the antennas can be matched to the often very different gain requirements of each band. Also, no high-gain vhf-uhf antenna made has a uhf section capable of the reception range of its vhf section. A separate array of high-gain uhf antennas with a low-noise preamplifier is necessary to match the reception capability of a good 10-foot vhf antenna (or 10-foot vhf section on a larger vhf-uhf antenna).

ISOLATION NETWORKS

The normal procedure in designing an all-channel antenna is to apply the vhf section to the uhf section in some manner, and connect the transmission line to the output terminals of the uhf section (which

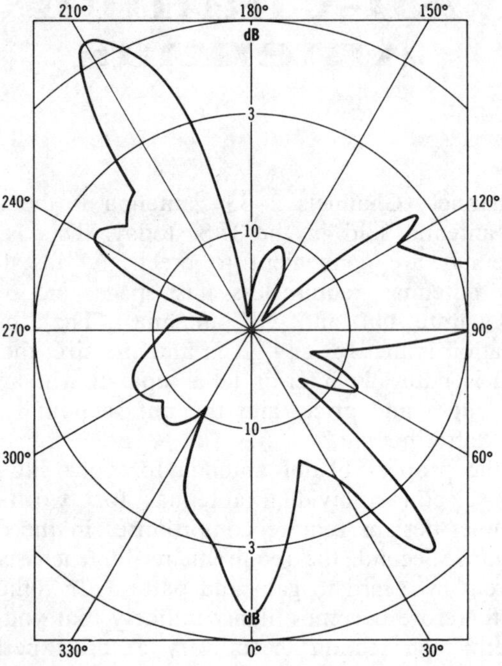

Fig. 4-1. Uhf polar pattern of vhf-uhf antenna without isolation network.

now carries both vhf and uhf signals). However, the vhf section of the antenna will pick up uhf signals also, nearly always from undesired directions. These undesired uhf signals effectively destroy the directionality of the uhf antenna, and make radical changes in its gain characteristics because of phase differences between the signal of the uhf antenna and the uhf signal of the vhf antenna. This is shown in the polar pattern of Fig. 4-1. The vhf-uhf antenna with which this pattern was made had no isolation network between its vhf and uhf sections. For proper performance, therefore, the output of the vhf section must be applied to the uhf section in some manner that eliminates the uhf component.

When the uhf section is a log-periodic antenna, the solution is simple. The output of the vhf section is applied to the rear terminals of the feeder lines of the log-periodic. As explained in Chapter 3, only signals *below* the uhf band (i.e., vhf signals) will reach the front of the uhf antenna, where the transmission line is attached. However, most vhf-uhf antennas use a corner-reflector yagi for the uhf section, which does not have an intrinsic blocking effect. This means that an external blocking network must be used to "choke off"

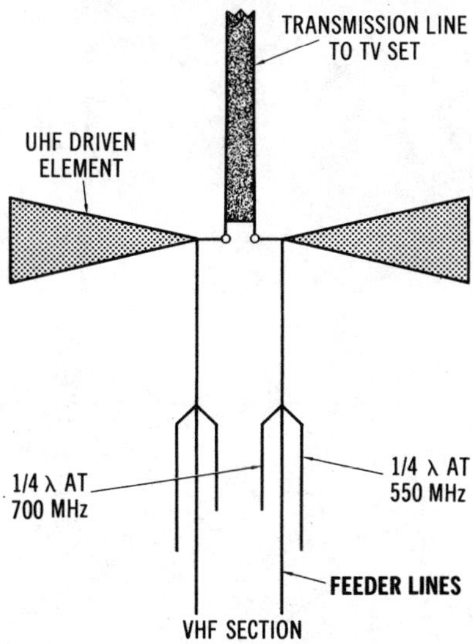

Fig. 4-2. Uhf isolation stubs.

the undesired uhf signals from the vhf section. The most popular (and probably most effective) technique is the one-quarter wavelength stub, shown in Fig. 4-2. These stubs are connected across the line connecting the vhf section to the uhf section. They short-circuit the lines to signals at frequencies where the stubs are one-quarter wavelength long. By making these stubs one-quarter wavelength long at uhf frequencies, uhf signals at that point on the line are shorted, while the vhf signals are not. This technique is used in Channel Master's Ultra-Hi Crossfire series. RCA antennas use two stubs of different length (per the illustration) for increased bandwidth, as does Winegard, Antenna Corp. of America, and Jerrold Electronics.

COMMERCIAL ANTENNAS

All of the vhf-uhf antenna models tested by the author were in the small to medium-size category. As was the case in Chapter 2, very small antennas (those having only two or three elements in their

(A) ES-270 antenna.

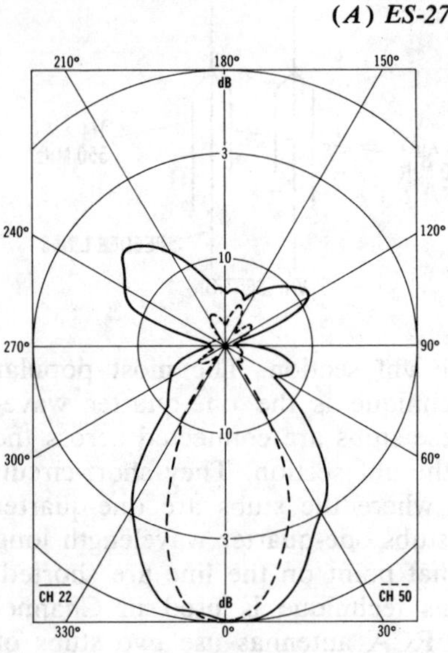

(B) Uhf polar patterns.

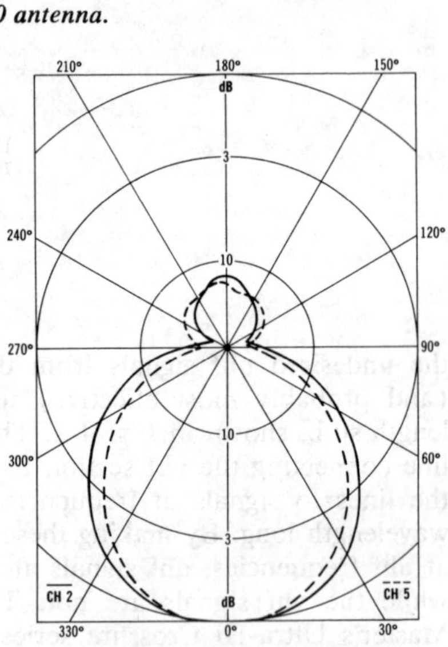

(C) Vhf-L polar patterns.

Fig. 4-3. Antenna Corp. of America

vhf section) are not included because of poor polar patterns and gain characteristics. Large (over 10 feet) antennas were neither tested nor covered in any way because of the reasons given at the introduction to this chapter.

Most of the vhf-uhf antennas selected for this chapter have vhf sections very similar to vhf-only antennas featured in prior chapters. Furthermore, most of the vhf-uhf antennas tested and/or recommended use corner-reflector yagis as their uhf section. This is because the corner-reflector design is very adaptable to the construction techniques used to make commercial antennas collapsible for shipping, and also provides fairly high gain without excessive boom length.

Strong- to Medium-Signal Antennas

The Antenna Corp. of America (A.C.A.) ES-270 shown in Fig. 4-3A has the longest boom in its price class. However, this is because the type of uhf section employed requires a lot of boom space. The vhf section is therefore about the same size as the others in its price class, containing three driven elements, a reflector, and two high-band directors. The rearmost director is spaced close to the first driven element, so it acts as a parasitic multimoding element. Another parasitic element near the second driven element completes the multimoding for this antenna.

The uhf section is simpler in design and construction than any of the others discussed in this chapter—a yagi with eight directors. The

(D) *Vhf-H polar patterns.*

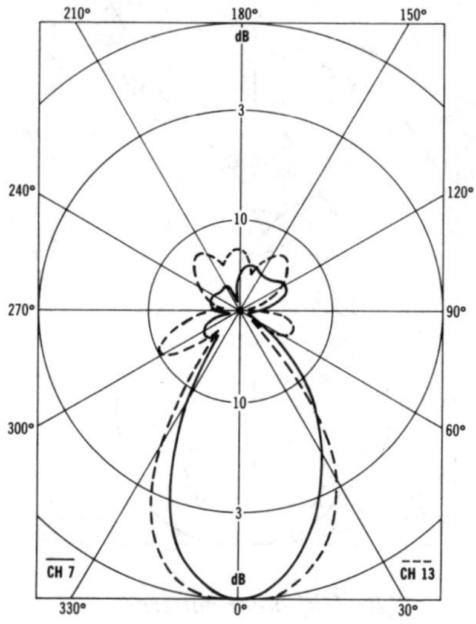

ES-270 antenna and polar patterns.

driven element is a three-quarter-wavelength folded dipole with dual-length isolation stubs. The frontmost vhf director acts as a uhf reflector, but because it is 1 to 1½ wavelengths long at uhf fre-

Courtesy Sony Corp. of America

(A) 4BG15 antenna.

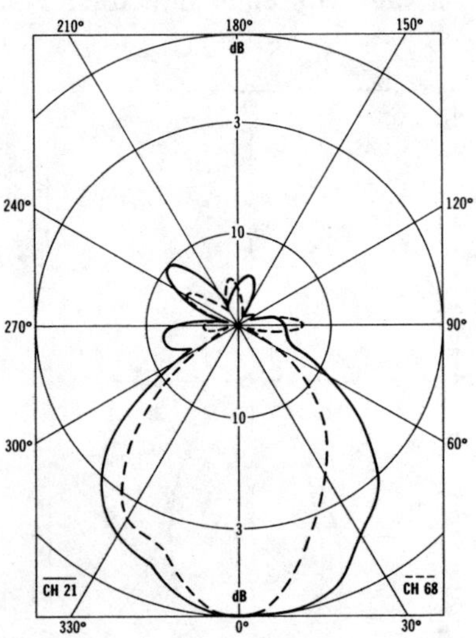

(B) Uhf polar patterns.

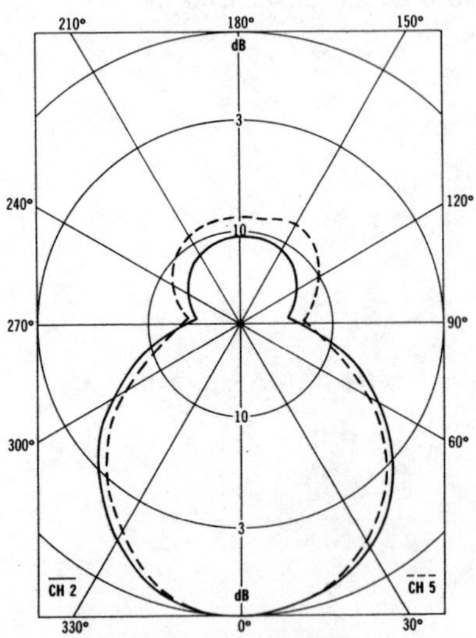

(C) Vhf-L polar patterns.

Fig. 4-4. RCA 4BG15 antenna

quencies, its performance varies considerably with frequency. As Fig. 4-3B shows, the performance on some channels is incredibly good, and on others only passable. Nevertheless, gain is higher over the midband channels than any of the others in its price class. Gain falls off rapidly past Channel 55 because the ES-270 directors are made quite long for highest gain on the most active portion of the uhf band (Channels 14–50). However, the directors do have score marks that allow them to be shortened if channels at the upper end of the band must be received. As was the case with all of the other antennas in this section, the polar patterns and gain measurements were made with the directors in the same condition as shipped (full length). Keep this fact in mind as you compare the high-end gain performances of these antennas. Note that high-end gain drops fastest on those antennas with the most directors when the directors are not shortened to improve the gain in this area.

The ES-270 has the best low-band polar patterns of the antennas in its price class (Fig. 4-3C), although the measured gain was slightly lower than the others. Therefore, you can expect excellent vhf reception to 20–25 miles, and Channel 25–55 reception to 15 miles with full-length directors.

For locations 10–20 miles from the transmitter, the RCA 4BG15 (Fig. 4-4A) is an excellent choice. The vhf section of this antenna is very similar to the RCA 3BG09 discussed in Chapter 2. The 4BG15 has three driven elements and a reflector. The centermost elements of the uhf corner reflector also act as a broadband vhf high-band director in the manner of the Winegard double director of Fig. 1-12A, and provide parasitic multimoding for the first driven ele-

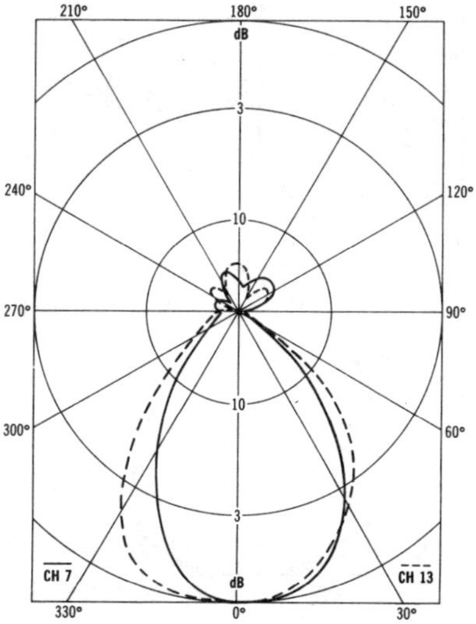

(*D*) *Vhf-H polar patterns.*

and polar patterns.

ment. The second driven element is multimoded via extension stubs. The result is vhf-H polar patterns (Fig. 4-4D) that are excellent for a small antenna. The uhf antenna section consists of a very small corner reflector with a bowtie driven element and two directors. The uhf polar patterns (Fig. 4-4B) are therefore fairly broad.

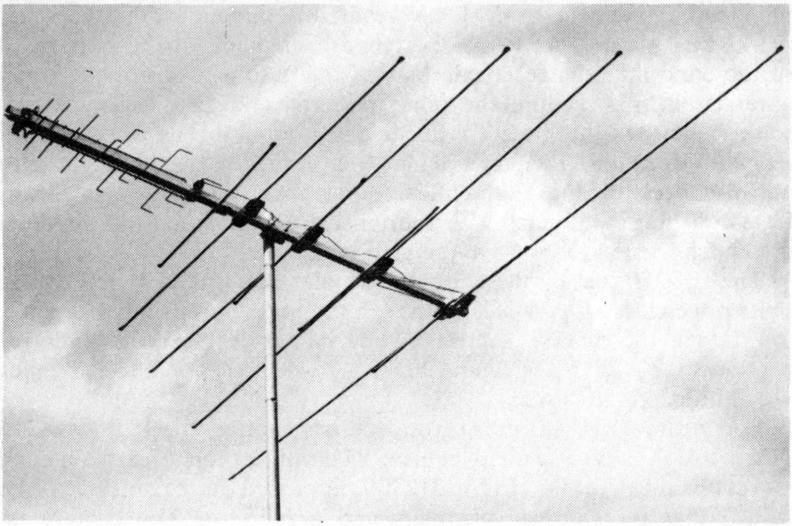

Courtesy Sony Corp. of America

(*A*) *CDX-650 antenna.*

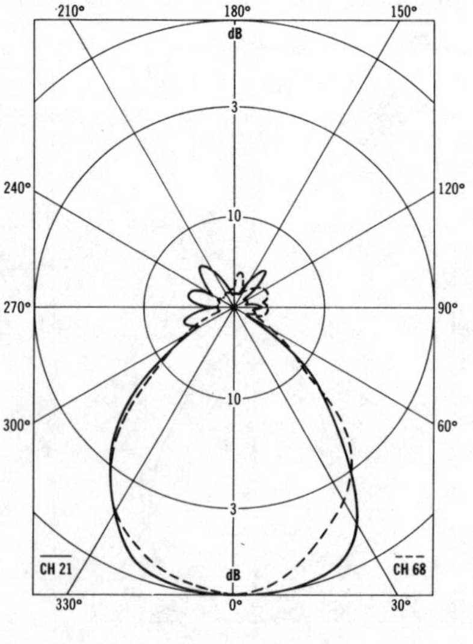

(*B*) *Uhf polar patterns.*

Fig. 4-5. Antennacraft CDX-650

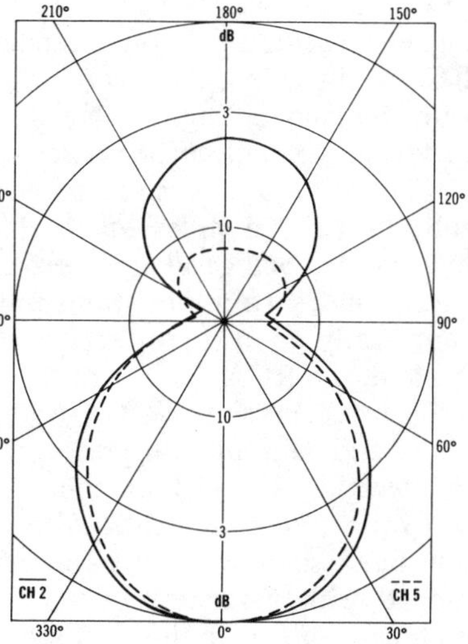

(*C*) *Vhf-L polar patterns.*

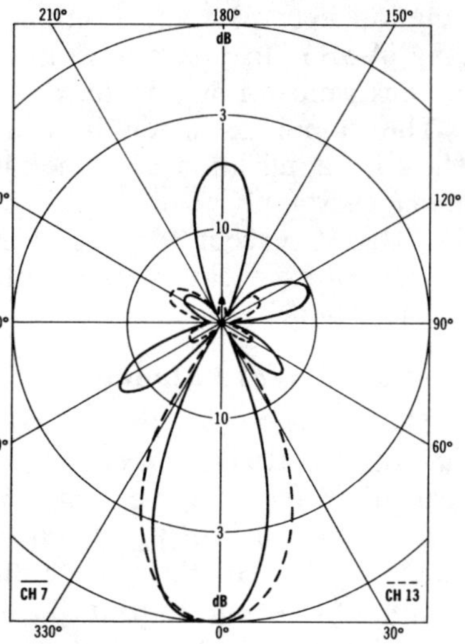

(*D*) *Vhf-H polar patterns.*

antenna and polar patterns.

According to the manufacturer's data, this antenna averages a smooth 2 dB gain on the vhf low band (nearly the same as the 3BG09), but only 2½–4½ dB on the vhf high band (less than the 3BG09). As was the case with the 3BG09, the 4BG15 had much higher measured gain on Channel 2 than most of the antennas tested. RCA's uhf gain curves show 5–8 dB gain over the active channels, with the gain falling off sharply at the low end of the band because of the very small corner reflector.

The 4BG15 has sufficient vhf gain to work well in the near suburbs (20–25 miles on the high band, and over 30 miles on the low band), but good uhf performance should not be expected past 10–15 miles from the transmitter, unless the receiving site is exceptionally good. If top uhf performance in the near suburbs is important, the RCA 4BG20 or 4BG26 is a better choice.

The Antennacraft CDX-650 (Fig. 4-5A) features excellent vhf high-band gain and uhf performance for a small (60-inch boom) and lightweight (1.6 kg) antenna. The vhf section of this antenna has just four driven elements, with the rearmost element shorted on the low band so it will act as a reflector. All four elements are multimoded with parasitic elements, so the high-band gain averaged as high as antennas in the next size and price category, with correspondingly narrow beamwidths (Fig. 4-5D).

The CDX-650 is the only one of the antennas tested that uses a log-periodic uhf section. The uhf feedthrough capability of the log periodic allows the vhf section's output to be directly connected to the rear end of the uhf-element feeder lines without the need of an isolation network. The result is extremely clean uhf polar patterns (Fig. 4-5B). In fact, the little CDX-650 had the highest uhf front-to-back ratios of any of the vhf-uhf antennas tested.

The comments made in Chapter 2 regarding the Antennacraft Mark-8's application as a mobile-home and recreational-vehicle antenna also apply to the CDX-650. Its range is about the same as the 4BG15 except for high-band stations. Here the CDX-650 is capable of excellent-quality pictures at 30–40 miles or so, depending on the channel.

Medium-Signal Antennas

The vhf section of the RCA 4BG20 (Fig. 4-6A) is identical with that of the 4BG15 except for an additional parasitic multimoding element in front of the first driven element. This results in higher average gain on the high band. RCA claims 4–6 dB high-band gain (same as the 3BG09); the author's measurements also show comparable improvement over the 4BG15. The most important difference, though, is in the uhf performance. The 4BG20 has a much larger corner reflector (which greatly improves the low-end gain) and four directors (which improve the high-end gain). These factors also significantly reduce the beamwidth at both ends of the uhf band (see Fig. 4-6B). The manufacturer's gain claims of 7½–9 dB over the active channels agree with the author's measurements, which showed up to 5 dB improvement over the 4BG15. The net result

(*A*) *4BG20 antenna.*

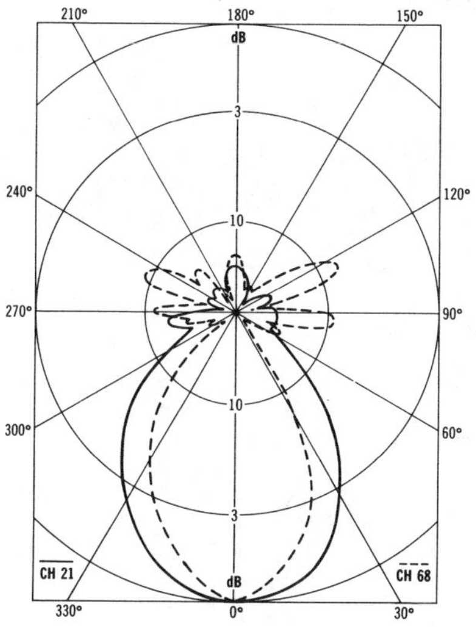

(*B*) *Vhf polar patterns.*

Fig. 4-6. RCA 4BG20 antenna and polar patterns.

is an antenna with the same fine vhf performance as the 3BG09, and decent uhf performance.

As shipped, the uhf directors on the 4BG20 (and other RCA models with four to six directors) are designed for maximum gain

Courtesy Sony Corp. of America

(A) 3676A antenna.

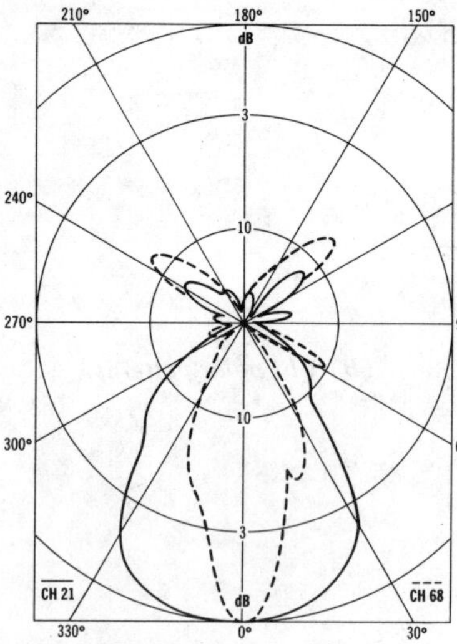

(B) Uhf polar patterns.

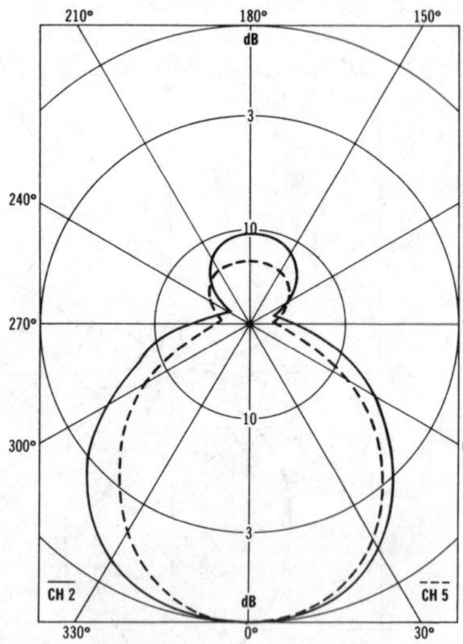

(C) Vhf-L polar patterns.

Fig. 4-7. Channel Master 3676A

over the most active channels. The gain of the 4BG20, for instance, starts to drop rapidly by Channel 70, so the antenna is nearly useless over the translator frequencies. However, snapping off the ends of the uhf directors (as shown on the instruction sheet accompanying the antenna) will peak the gain for fine performance in the translator band.

The 4BG20 has sufficient vhf gain to work well in the suburbs, and sufficient uhf gain to work well to 20 miles or so from the transmitter. If comparable vhf-uhf performance is needed, the ideal antenna is the RCA 4BG26. This has a vhf section identical with the 4BG20, but a very large uhf antenna section. In fact, the 38-inch-high, 23-inch-wide corner reflector on the 4BG26 is much larger than the usual uhf-only corner reflector, and as large as is commonly found on vhf-uhf antennas of any size.

Channel Master's 3676A (Fig. 4-7A) is the most ruggedly constructed vhf-uhf antenna tested by the author, yet has a surprisingly low price. Reinforcing sleeves extend out from the long elements of the vhf section: very heavy metal stampings are used for the uhf directors; ⅜-inch bolts clamp the boom to the tv mast. A bonus feature is that the Model 0012 band separator supplied with this antenna is the best 300-ohm band separator the author has ever seen.

The vhf section of the 3676A has three driven elements, a reflector, and a pair of collinear high-band directors. Bimoding is by parasitic elements (one per driven element), so the high-band gain (a smooth 7 dB according to Channel Master) is very high for the number of driven elements. The measured low-band gain averaged

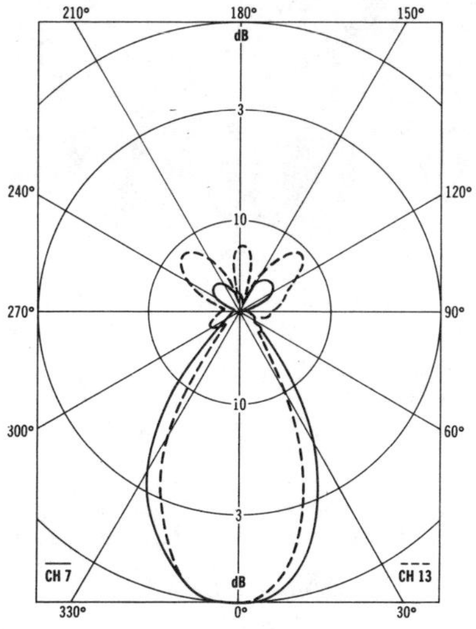

(*D*) *Vhf-H polar patterns.*

antenna and polar patterns.

out about the same as the other antennas already discussed. This is to be expected since the 3676A and RCA's 4BG15, 4BG20, and 4BG26 have the same element configuration on the low band.

The uhf section consists of a corner reflector with a three-quarter-wavelength folded dipole driven element and four directors. The

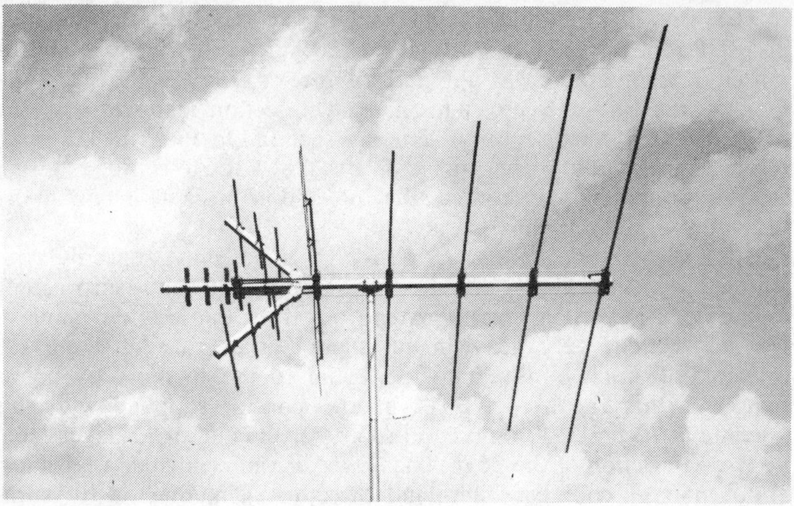

Courtesy Sony Corp. of America

(A) VU-932S antenna.

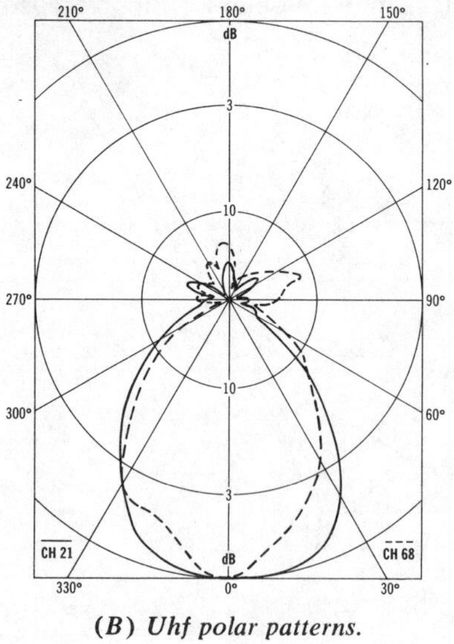

(B) Uhf polar patterns.

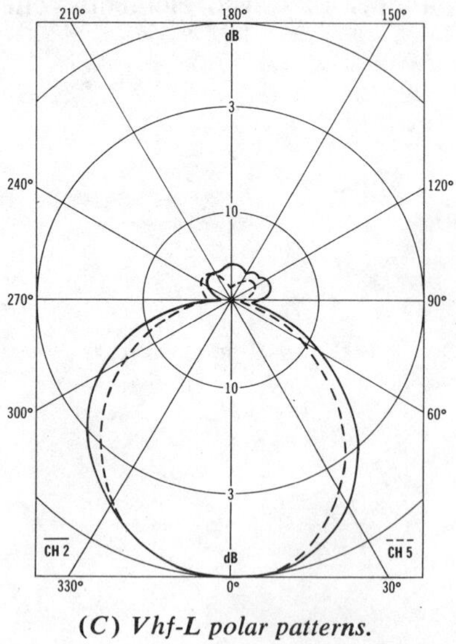

(C) Vhf-L polar patterns.

Fig. 4-8. Jerrold VU-932S

claimed gain of 7–11 dB over the active channels with peak gain around Channel 55 is totally consistent with that measured by the author. The 3676A had the highest measured uhf gain of the vhf-uhf antennas tested.

Like the antennas previously covered, the 3676A will provide excellent vhf-L reception to 30 miles or so, but can handle 40 miles easily on high-band signals. The uhf range is 20–25 miles, depending on the channel.

The 3675A is the next step up in Channel Master's Crossfire line, having an additional vhf driven element and two extra uhf directors. The 3675's uhf and high-band vhf gain figures are only a ½ dB better than the 3676A, but the low-band performance is much better, especially across the fm band. Less than ¾ dB variation across the full fm band makes this antenna a front-runner when fm reception is desired from your tv antenna.

The Jerrold VU-932S (Fig. 4-8A) is a well-made antenna featuring unusually good performance for a small boom length (69 inches). Its vhf section has five driven elements, but the last one is shorted so it acts as a reflector. There is one distinct high-band director, and a second one formed by the end rods of the uhf corner reflector. Multimoding is by Jerrold's unique method; the result is very high gain (7–8½ dB) on the high-band channels. Since this antenna has a greater number of elements working on the vhf-L channels than the vhf-uhf antennas already covered, its vhf-L polar patterns (Fig.

(D) Vhf-H polar patterns.

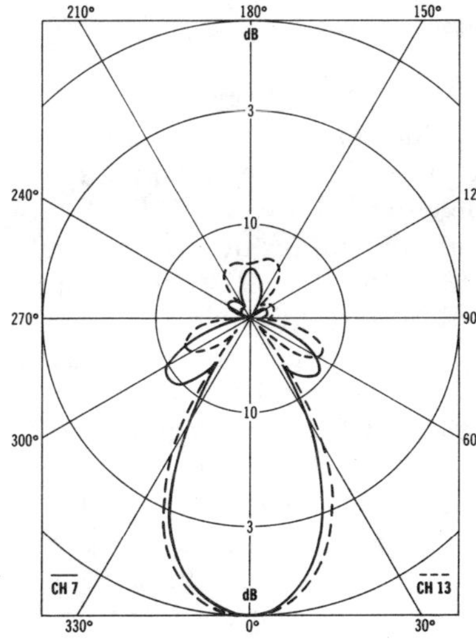

antenna and polar patterns.

4-8C) show much higher front-to-back ratios. The important thing about the low-band pattern, however, is the very deep side nulls. If fm interference on Channel 6 or cochannel tv intereference occurs, the antenna's orientation can often be adjusted for minimum pickup of the interfering signal by dropping it into a side null. However, if fm reception is desired, the VU-932S is also a good choice; its low-band gain rolls off so gradually that 3–4 dB gain is available over the fm band.

The measured uhf gain of the VU-932 averaged out about the same as the RCA 4BG20 (which has a similarly sized corner-reflector), so Jerrold gain claims of 7½–8½ dB over the active channels agree with everyone's findings. The polar patterns (Fig. 4-8B) were fairly nice, but the really interesting characteristic of this section is that the gain doesn't drop sharply at the translator frequencies. The gain held up well to Channel 75 (the highest channel measured by the author) using the full director lengths. The gain can be boosted further on the translator channels (70–83) by breaking off the ends of the three butterfly directors. For higher gain over Channels 14–70, the Jerrold Model VU-8PZ director array can be installed.

The various factors discussed mean the VU-932S is a compact antenna capable of excellent vhf performance to 30 miles on the low band, and 45 miles on the high band. On uhf the estimated range for a Grade A picture is 20 miles; this and the good patterns suit this antenna to metropolitan and suburban areas where ghosts are a problem.

Weak-Signal Antennas

For the reasons mentioned at the beginning of this chapter, very large vhf-uhf antennas are neither recommended nor covered in this chapter. The Jerrold VU-934S (Fig. 4-9A) is the largest vhf-uhf antenna the author deems practical.

The 10-foot-long VU-934S has eight driven vhf elements multimoded by Jerrold's unique method. As is the case with the VU- and VIP-series antennas, the rearmost driven element is shorted (to act as a reflector) and the frontmost has Jerrold's "trombone" capacitive element for best Channel 6/Channel 7 performances. However, the VU-934S has a few additional features. First, high-band vhf directors are arranged in collinear fashion with the uhf driven element. Second, high/low vhf directors are used as end elements in the uhf corner reflector. Thus, the same elements function as vhf-L directors, vhf-H directors, and uhf reflectors, with the effect that gain is boosted greatly over the upper halves of both vhf bands, and over the bottom half of the uhf band. Six butterfly directors boost uhf gain from the middle to the high end of the band.

The net result is an antenna with 6–7 dB vhf-L gain, 11–12 dB vhf-H gain, and 9–10½ dB uhf gain, *as shipped*. Like the vhf-only VIP-303, the VU-934S has high/low directors designed to reduce fm pickup. As the solid lines in the vhf gain charts in Fig. 4-10 show, fm gain is far below the vhf-L gain. When the high/low director ends are broken off, a little gain is lost at the upper ends of each vhf band, but 6–6½ dB gain across the fm band is acquired (see dashed lines in the vhf gain curves). In fact, the VU-934S is one of the few tv antennas available that are capable of fm gain of this magnitude. These gain figures mean excellent vhf picture quality at 50–55 miles, and comparable uhf reception to 25 miles. High-quality uhf reception can be extended another 5 miles or so by adding the 8PZ director array. The dashed line in the uhf gain chart shows the resulting frequency response.

Measured Performance Comparison

In Table 4-1 the vhf-uhf antennas tested by the author are compared as to physical characteristics and measured electrical per-

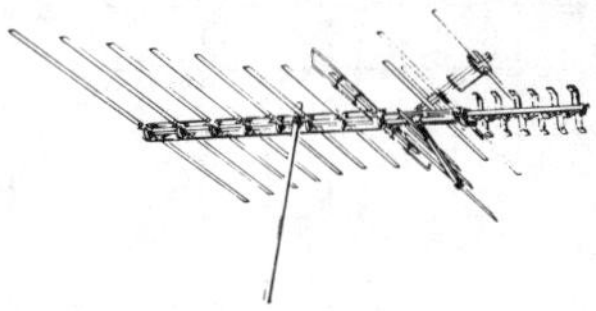

(*A*) *VU-934S antenna.*

Courtesy Sony Corp. of America

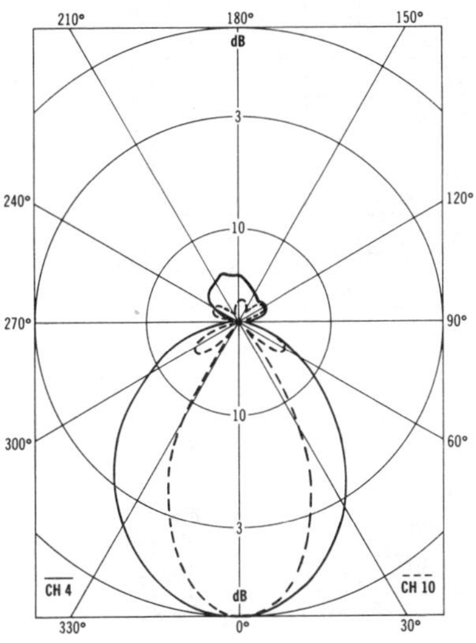

(*B*) *Vhf polar patterns.*

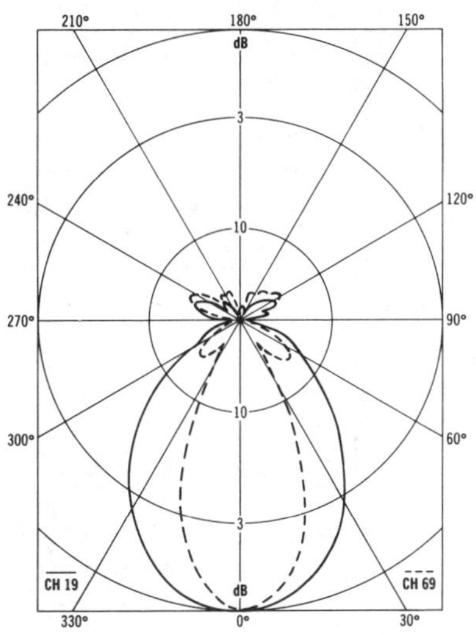

(*C*) *Uhf polar patterns.*

Fig. 4-9. Jerrold VU-934S antenna and polar patterns.

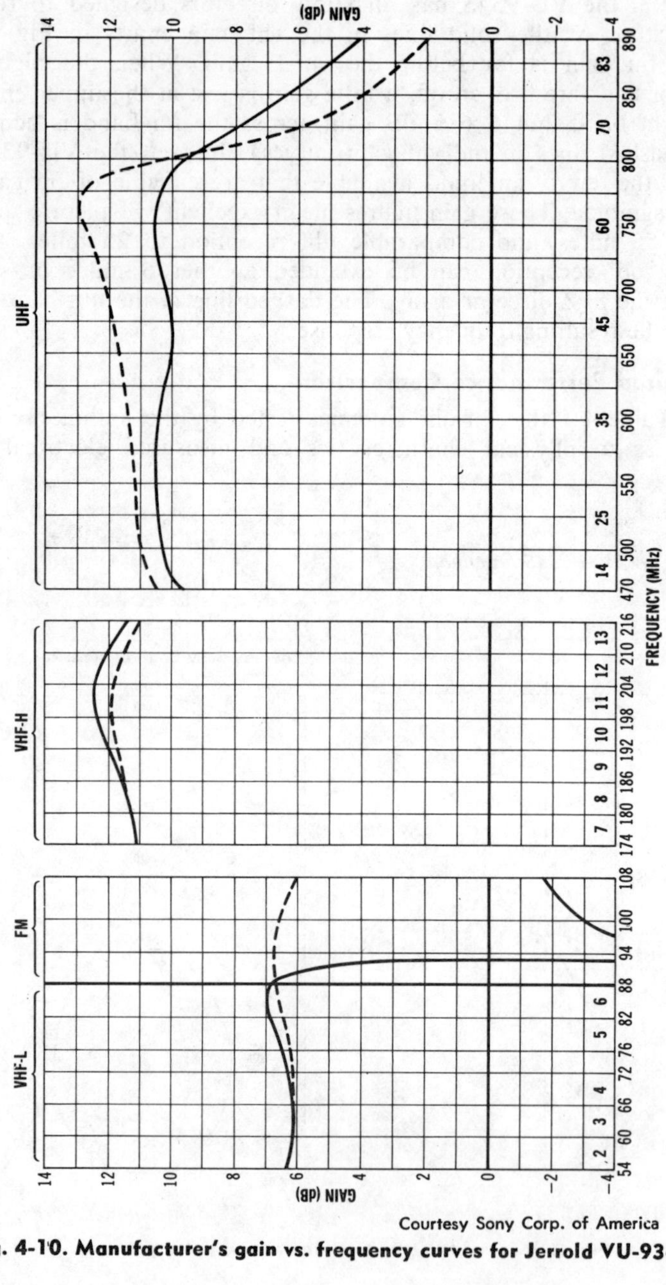

Courtesy Sony Corp. of America

Fig. 4-10. Manufacturer's gain vs. frequency curves for Jerrold VU-934S.

Table 4-1. Comparison of Measured Antenna Performance

Antenna	Boom Length (in)	Wt. (kg)	Prices ($) List	Prices ($) Dlr	Gain (dB) 2	4	5	7	9	11	13	Beamwidth (°) 2	5	7	13	Front-to-Back Ratio (dB) 2	5	7	13
A.C.A. ES-270	84	1.7	33	20	2.0	2.0	2.1	3.6	4.9	6.7	5.9	83	82	42	53	12	12	15	12
RCA 4BG15	68	1.9	33	20	4.6	2.7	3.0	4.0	5.0	6.5	6.2	90	83	52	67	10	9	17	16
Antennacraft CDX-650	61	1.6	33	20	2.6	3.0	2.7	4.0	6.5	7.5	7.4	87	79	30	41	4.4	12	5.7	14
RCA 4BG20	80	2.0	38	23	4.9	2.0	2.5	5.0	5.7	6.8	6.4	91	81	61	71	10	12	14	18
Ch. Master 3676A	77	2.3	36	22	2.8	3.3	2.5	6.5	7.2	6.5	7.0	92	71	49	40	10	14	16	11
Jerrold VU-932S	69	2.1	45	27	1.4	3.0	3.5	7.4	7.6	8.3	7.0	88	76	42	47	17	18	16	14

Antenna	Type	Gain (dB) 20	25	30	35	40	45	50	55	60	65	70	75	Beamwidth (°) 21	50	68	Front-to-Back Ratio (dB) 21	50	68
A.C.A. ES-270	Yagi	4.1	5.6	7.0	7.9	7.8	10.4	10.7	9.3	2.7	0.2			56	35		6	16	
RCA 4BG15	CRY	4.8	5.5	7.1	6.6	7.8	7.0	7.2	7.5	5.1	4.6	5.5	2.7	83		56	10		15
Antennacraft CDX-650	LP	4.7	5.8	8.2	7.4	4.9	3.3	7.0	5.6	7.7	5.9	3.6	2.2	67		67	15		18
RCA 4BG20	CRY	8.1	8.3	7.2	8.4	8.7	7.8	8.2	7.3	10.1	8.3	6.2	−4.5	64		46	16		8.4
Ch. Master 3676A	CRY	8.2	8.4	8.5	8.5	8.8	9.3	10.1	9.6	11.3	7.4	6	−2.5	66		21	12		8
Jerrold VU 932S	CRY	6.3	6.8	7.7	8.6	8.7	8.8	8.9	8.9	8.7	8.5	7.8	6.6	67		55	16		12

CRY = Corner-reflector yagi LP = Log-periodic

formance. The same comments made in Chapter 2 in regard to price, gain figures, front-to-back ratio and beamwidth apply also to the corresponding data in this table.

The figures in the "Gain (dB)" columns are the author's attempt to present measured relative gain data in the most useful manner by *approximating* absolute gain. This was done by using a composite of manufacturer's gain data to set figures for one antenna, and then calculating the other antenna gain figures from that by means of the measured relative gain data.

The vhf low-band output of vhf-uhf antennas is greatly influenced by the characteristics of the test site, because the distance between the antenna and its surroundings (roof, wires, structures) is small in terms of wavelength. In some cases the antenna output is greatly enhanced on a particular channel (e.g., the RCA 4BG15 and 4BG20 on Channel 2), while in other cases the antenna output is decreased (e.g., Jerrold VU-932S). If the antennas are mounted at roof-to-antenna distances much greater than the 12 feet used for the author's measurements, the relative gain may be different.

CHAPTER **5**

Transmission Lines

A good signal at the antenna has to conveyed to the tv set without significant degeneration to be of any use. The transmission line is the first (and often only) link in the "pipeline" that accomplishes this.

There are many types of transmission lines available, as you will find out in this chapter. Though each has some unique feaures, all have two characteristics that are keystones in the matter of line selection: impedance and attenuation.

Television transmission lines are manufactured for either 300 ohms impedance (balanced line) or 75 ohms impedance (coaxial cable). As most (but not all) tv receivers and antennas are designed for direct interface at 300 ohms, lines of this impedance have the advantage of direct connection to antenna and tv set. However, a simple device called a *balun* can effect an impedance change of 300 to 75 ohms, or 75 to 300 ohms, so the advantages of coaxial cable can be enjoyed at small cost.

All transmission lines reduce the strength of the tv signal on the way to the set. This *attenuation* or loss is directly proportional to the length of the line, so is commonly expressed in *decibels per 100 feet*. Attenuation also varies with frequency, so each attenuation figure is always specified at a particular frequency. This means that a 200-foot run of line with 6 dB loss per 100 feet at 500 MHz will attenuate the signal by 12 dB at that frequency. Similarly, a 35-foot length of the same line will cause a signal loss of about 2 dB.

BALANCED LINE

Balanced line consists of two parallel conductors separated by air or a solid insulating material. When the dielectric (the substance between the conductors) is air, the balanced line is known as *open line.* When the dielectric is solid polyethylene, the balanced line is known as *twinlead.* In either case, the conductor spacing is adjusted to yield 300 ohms impedance for the conductor size used.

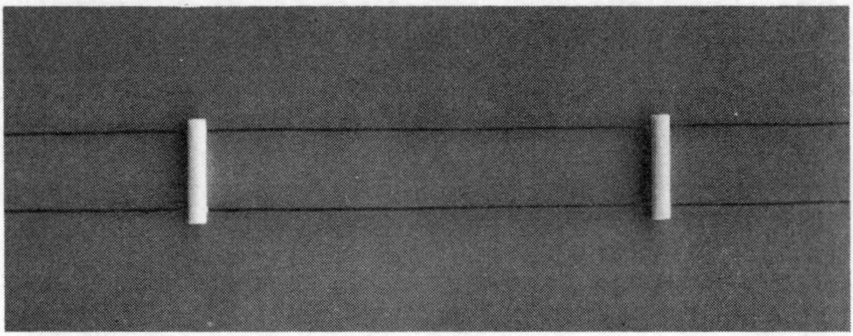

Fig. 5-1. Saxton 2502 open line.

Open Line

Open line (Fig. 5-1) has potentially lower losses than any other type of line. The dry-weather losses of 300-ohm open line with No. 18 conductors (Saxton 2502 for example) are less than 2 dB per 100 ft at 900 MHz, and this increases by less than 10 percent when the line is wet. Similarly, it is only slightly degraded by aging in an industrial (polluted) atmosphere, and will last indefinitely. Unfortunately, open line suffers from the same proximity effects as twinlead (described in the next section), and requires even greater care during installation. Maintaining the proper separation between conductors during a compound curve is difficult; doing this when the line connects to a rotated antenna is next to impossible. Moreover, insulators capable of spacing the line at least 10 inches from the antenna mast and building are needed to fully realize the low theoretical losses of open line. Unfortunately, insulators suitable for this task are not commercially available. The chief application of open line is in rural areas for the extremely long (over 1000 feet) straight run from a hilltop antenna to a home in the valley. The line can be nailed to tree trunks or poles along the way with commercially available insulators. Thus the line is clear of surrounding objects for over 95 percent of its run, and its actual losses will be close to the theoretical.

Twinlead

Twinlead is available in many forms, some of which are shown in Fig. 5-2. Standard 300-ohm twinlead, having No. 22 stranded conductors and flat solid-polyethylene dielectric, is the cheapest tv transmission line available. When new and dry, and installed in an ideal (theoretical) environment, standard twinlead has very low loss at any frequency. Premium versions of this line, using No. 20 conductors and somewhat heavier construction, have even lower losses, second only to open line. This performance, however, is very difficult to obtain in typical, or even practical, installations. With the exception of shielded twinlead, all types of twinlead and open line are very sensitive to nearby metallic objects and nonmetallic lossy structures. Insulators must be used where needed along the line to keep it at least 4 inches from nonmetallic surfaces and 10 inches from metal surfaces (such as rain gutters, aluminum storm windows, drain pipes, antenna masts, etc.). Even then, the insulators themselves cause impedance irregularities. Metal surfaces nearer to one conductor than the other unbalance the line, causing signal pickup that appears on the tv screen as a ghost. Detailed "do's" and "dont's" about installing twinlead are given in Chapter 10.

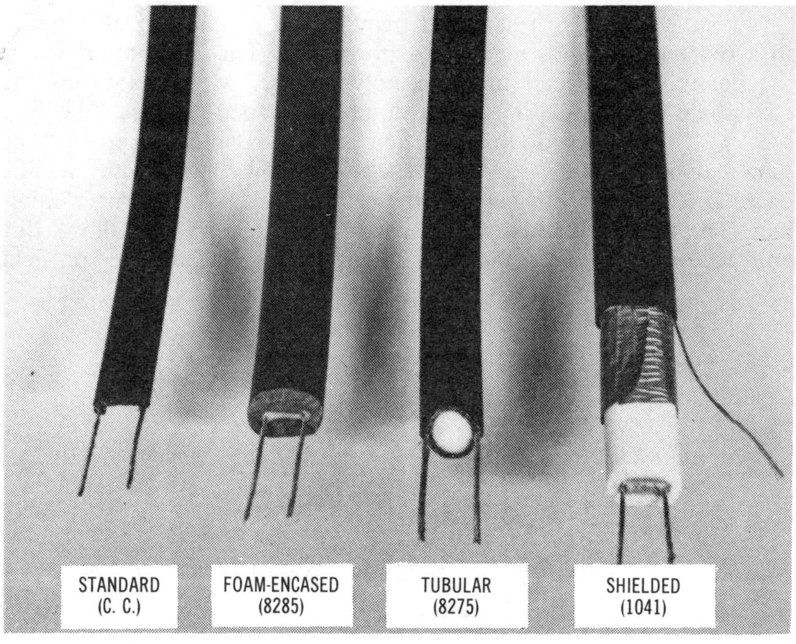

Fig. 5-2. Various types of 300-ohm twinlead.

The biggest problem with standard twinlead (and to a lesser extent the more exotic twinleads described in the next section) is the effect of water (be it rain or ice) on the loss characteristics of the line. The figures in Table 5-1 show that rain causes an enormous

Table 5-1. Dry- vs. Rainy-Weather Attenuation for Balanced Lines

Type of Line	Cond. Size	dB Attenuation per 100 Feet: 100 MHz Dry	100 MHz Rainy	500 MHz Dry	500 MHz Rainy	900 MHz Dry	900 MHz Rainy
Open Line	18	0.4	1.2	0.8	2.3	1.8	4.9
Standard Twinlead	22	1.4	7.5	3.8	20	5.6	30
Foam-Encased Twinlead	22	1.4	2.0	3.8	5.2	5.6	7.5
Tubular-Foam Twinlead	20	1.0	2.6	3.0	6.6	4.3	9.0
Shielded Twinlead	26	3.6		8.4		11.5	
Shielded Twinlead	22	2.2		5.5		7.7	

increase in line losses. The losses are even higher with radial ice loading. Either way, the wet attenuation figures of standard twinlead are staggering, particularly at uhf frequencies.

Foam Twinlead

Several variations of standard twinlead have arisen in an attempt to avoid the problems mentioned previously. The basic approach is to encase the twinlead in foam polyethylene. Most manufacturers offer foam twinlead with a flat or dumbbell-shaped cross section (Fig. 5-3A). These lines do have better adverse-condition performance than standard twinlead, and are less affected by contact with metallic objects. However, with most of these lines the improvement is minimal. The narrow insulation between the conductors still allows dirt and water to intrude in the most intense portion of the electric field

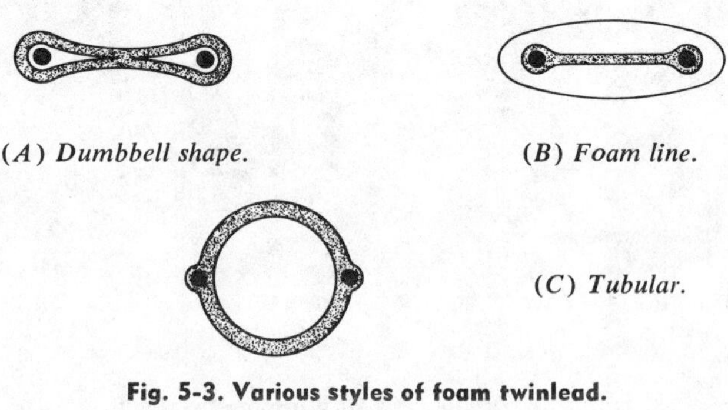

(*A*) *Dumbbell shape.*

(*B*) *Foam line.*

(*C*) *Tubular.*

Fig. 5-3. Various styles of foam twinlead.

(between the conductors). A much better approach is used in Belden 8285 foam line (Fig. 5-3B). Standard twinlead is encased within a *thick* wrapper of foam polyethylene that functions as a physical barrier to keep dirt, moisture, and objects far from the conductors *and* the most intense part of the electric field. The foam wrapper on this cable is thick enough to be highly effective; 8285 is the least sensitive of the unshielded twinleads to contact with metallic objects, and its wet attenuation figures are excellent for an unshielded line. Even at uhf frequencies, line losses increase only modestly when wet.

For reception sites very far from the transmitter, and which are also located in very dry climates (a vacation home in the Arizona desert for example), Belden 8275 tubular twinlead (Fig. 5-3C) is a potentially better choice. This line has No. 20 conductors to minimize skin losses, and is foam filled to minimize dielectric loss. The foam filling also prevents occasional moisture from entering the line (as is the case with hollow-core tubular twinlead). These points especially recommend Belden 8275 for uhf installations in dry climates; they result in line losses less than those of standard flat twinlead. The round cross section holds dust, pollutants, and moisture away from the most intense parts of the electric field, and also gives the line better wind and flexion characteristics than flat lines. However, greater care must be taken with this line to avoid contact with metallic objects than with 8285.

Belden 8285 and 8275 cost about twice as much as high-grade standard twinlead. Both are available in precut lengths of 50, 75, and 100 feet with terminals installed. The actual appearance of these lines is shown in Fig. 5-2.

Shielded Twinlead

Shielded twinlead (Fig. 5-2) consists of flat twinlead encased in a large amount of foam polyethylene, and wrapped in aluminized-Mylar foil and an outer jacket. The foam filling provides sufficient separation between the foil and line to prevent excessive attenuation by the foil. The foil shield makes the line immune to the disturbing effects of the outside world. A drain wire permits grounding the shield (if desired). The result is a balanced line that has considerably greater dry attenuation than standard twinlead, but whose attenuation and impedance characteristics are not affected by rain, dirt, age, or surrounding objects. Shielded twinlead can be taped to antenna masts and rain gutters, run through walls, etc.

These good features are obtained at no small cost. The lowest-attenuation shielded twinlead is nearly twice as costly and bulky as coaxial cable with similar attenuation characteristics, as will be shown in a following section. However, shielded twinlead offers the greatest overall immunity to noise pickup of any type of transmission line. The foil shield gives it nearly as much immunity to the electric-field component of radio-frequency and electromagnetic forms of interference as foil-and-braid coax. Because the signal conductors are balanced to ground and the shield is not part of the signal circuit, shielded transmission line also has high immunity to the magnetic-field component of radio-frequency and electromagnetic interference. The shield prevents unbalance due to proximity effects, so any magnetic-field pickup is due to unbalance caused by antenna asymmetry or unbalance in the tv-receiver input circuit.

Shielded twinleads also offer convenience and uncomplicated installation, in that the 300-ohm balanced lines offer a direct match to both antenna and tv set. Also, both low-attenuation (Saxton 1041 and Belden 8290) and small-diameter (Belden 9090) shielded lines are available in precut lengths of 50, 75, and 100 feet with terminals installed, a further convenience.

COAXIAL CABLE

Coaxial cable (or *coax*) consists of a thin central conductor within a larger cylindrical outer conductor. The central conductor is held equidistant from the walls of the outer conductor by an insulating material (dielectric). The inside diameter of the outer conductor (called the *shield*) is adjusted to yield 75 ohms impedance for the inner conductor size and dielectric material used.

Unlike balanced lines, coaxial cable is operated with the outer conductor at ground potential, hence coaxial cable is also known as "unbalanced line." This mode of operation has many advantages. The outer conductor does not have to be insulated from the rest of the world, so coaxial cable can be taped to antenna masts and rain gutters, run through walls, etc. The electric field between the two conductors is entirely within the cable, so the cable's attenuation and impedance characteristics are not affected by rain, dirt, age, or surrounding objects. Moreover, the grounded shield makes the cable immune to noise pickup and interference.

A variety of coaxial cable types used in home tv and small matv installations are shown in Fig. 5-4. They are arranged in order of size: the smallest at the top, the largest at the bottom. The model numbers on the cable jackets correspond to those mentioned in the following paragraphs.

Standard tv coax (Belden 8421 for example) is called type RG59U. This type of cable has a copper-braid shield, No. 22 center conductor, and solid-polyethylene dielectric. While this is perfectly adequate for vhf installations in strong- and medium-signal areas, and uhf installations in strong-signal areas, there are several less-expensive RG59-sized cables now available that have far superior performance. These cables have foam-polyethylene dielectric and No. 20 center conductors. These two factors result in a cable with about 25 percent less loss than standard RG59U. The newest and *least expensive* cables of this type have an aluminized-mylar foil shield instead of a copper braid. To make connection to the shield, four wires (called "drain" wires) are spirally wound around the foil (9234), or the foil is covered by a loosely woven aluminum braid (9275). The foil shield provides far better shielding effectiveness than

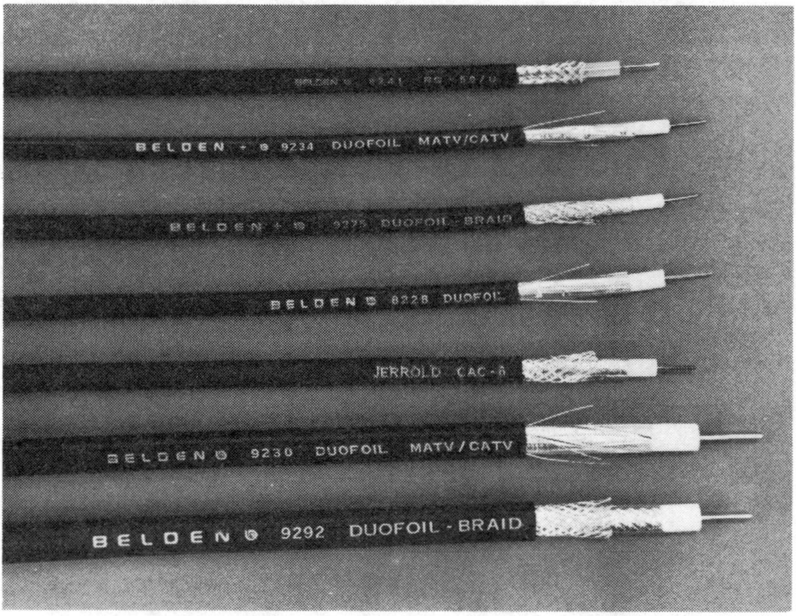

Fig. 5-4. Sizes and types of coaxial cable commonly used in tv and matv installations.

a braided shield, so this type of cable is especially recommended where interference is a problem. The foil-and-drain wire combination is the least expensive style but connector installation poses certain problems at times. The foil-and-aluminum braid combination costs slightly more, but has many advantages and is well worth the extra pennies. The aluminum overbraid provides mechanical protection to the foil during flexion, prevents radio-frequency leakage at the connector junction, and makes it easier to install connectors. Cables with the aluminum overbraid are also slightly larger in diameter than their drain-wire counterparts.

In certain foil-shield cables the foil is bonded to the dielectric, so the foil cannot accidentally be pushed down into the cable when the connector is installed. This type of cable permits exceptionally easy connector installation.

RG59-sized, foam-dielectric, foil-and-braid cables that performed better than the manufacturers' specifications in the author's measurements, and which also exhibited excellent mechanical quality, include Belden 9275 and 9282 (bonded foil), Jerrold CAC-59, and Winegard CL-2700 (bonded foil). While these cables all are available in 500- and 1000-foot spools for antenna installers, servicemen, and other bulk users, the Winegard CX series (Fig. 5-5) and Antenna Corp. of America HDW series of foil-and-braid foam cables are available in precut lengths of 25, 50, 75, and 100 feet with connectors and a rubber boot (for the antenna balun) installed! These prefabricated cables are made from the same type cable as Winegard CL-2700 and Belden 9282.

RG6U-type cable with foam-polyethylene dielectric and aluminized-Mylar foil shield should be used for long cable runs at uhf frequencies. This type of cable has a No. 18 center conductor for lower loss, but is still small enough for complete practicality in consumer applications. Although RG6-type cables with drain-wire construction (8228) are available, the recommended RG6U-type cables have an aluminum overbraid for easy connector installation. Low-priced cables having this construction and whose measured performance equaled or bet-

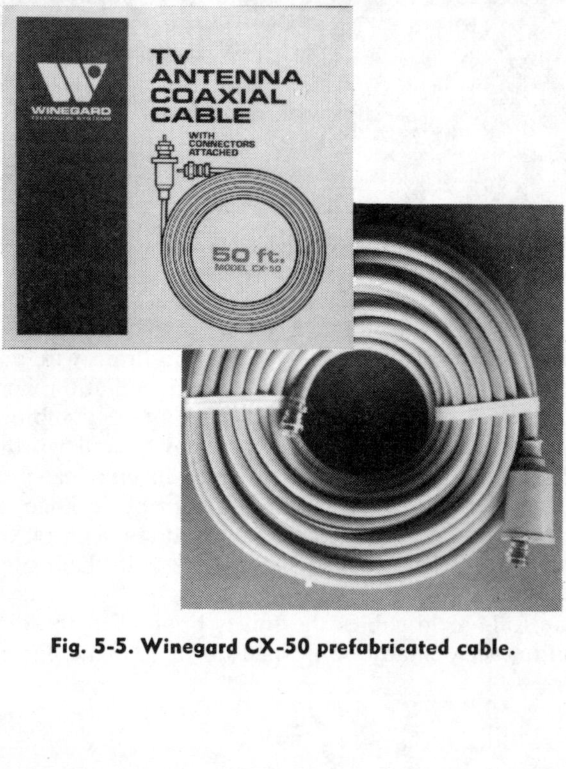

Fig. 5-5. Winegard CX-50 prefabricated cable.

tered their published specifications are Jerrold CAC-6, Winegard CL-2800, Finco CX-286 and Belden 9283 (bonded foil). Belden 9248 is a more expensive cable that is similar except for its overbraid. The overbraid on this cable is somewhat heavier (61 percent coverage vs. 40 percent coverage for the others mentioned) and solderable! This feature is useful for cases where the cable must be soldered to a chassis, or where special types of connectors must be installed.

RG11-sized cable is the largest coaxial cable that is usable in home and small matv installations. This cable has a No. 14 center conductor, so its losses are extremely low, in the vicinity of twinlead losses. It is available with foil-and-drain wire construction (Belden 9230) and foil-and-braid construction (Belden 9292 and Jerrold CAC-11). Because of its comparative stiffness and the fact that its large center conductor will not mate with some matv components, the use of RG11-sized cable is generally restricted to the main cable run for critical uhf installations. Cost is also a factor in its limited consumer applications; it costs about twice as much as RG6U-sized cables. However, it does offer a practical alternative when a mast-mounted uhf preamplifier cannot be used.

COMPARISON OF TRANSMISSION LINES

Now that the major physical and electrical characteristics of the various transmission lines have been discussed, we can determine what type of line is best for specific antenna installations.

During the discussion of applications, refer to Table 5-2 to see how the type of line under discussion compares to the other types of line. This table compares the physical and financial factors, and electrical performance of a variety of transmission lines. The loss characteristics are given in terms of decibel loss per 100 feet at four key frequencies. For uniform price and attenuation comparisons, all data is based on a single manufacturer (Belden) whenever possible. Belden numbers are also used for convenience or identification in the table and text. After selecting the transmission line for your application, refer back to the preceding sections of this chapter for the model numbers of other recommended brands of the selected line type.

The data in Table 5-2 was taken from the manufacturers' literature. The accuracy (and conservatism) of the coax attenuation figures was verified by the author's own measurements. In every case the measured performance equaled or bettered the rated performance, both for the model numbers listed in this table, and the other brands mentioned earlier in this chapter. The prices shown were based on 1000-foot spools at mid-1976 prices. They are for *comparison purposes only;* smaller lengths and packages of precut line will be much higher.

Table 5-2. Comparison of Various Types of Transmission Line

Type of Line	Cond. Size	Width or OD	Cost (per ft)	Belden Number	dB Attenuation per 100 Feet 100	200	500	900 (MHz)
Open Line	18	.500	5.2¢	2502*	0.4	0.7	1.2	1.8
Standard Twinlead	22	.400	2.5	1554*	1.4	2.2	3.8	5.6
Standard Twinlead	20	.400	3.0	8225	1.1	1.7	3.1	4.5
Foam-Encased Twinlead	22	.468	6.3	8285	1.4	2.2	3.8	5.6
Tubular Foam Twinlead	20	.400	6.1	8275	1.0	1.6	3.0	4.3
Shielded Twinlead	26	.345	9.4	9090	3.6	5.2	8.4	11.5
Shielded Twinlead	22	.515	13.2	8290	2.2	3.4	5.5	7.7
RG59U Solid Coax, Copper Braid	22	.242	7.1	8241	3.4	4.9	9.1	13.9
RG59 Foam Coax, Foil & Wires	20	.225	5.4	9234	2.7	3.8	6.2	8.6
RG59 Foam Coax, Foil & Braid	20	.242	5.1	9275	2.6	3.8	6.2	8.4
RG6 Foam Coax, Foil & Wires	18	.242	5.7	8228	2.1	3.1	5.0	6.9
RG6 Foam Coax, Foil & Braid	18	.273	6.9	9283	2.1	3.1	5.0	6.9
RG11 Foam Coax, Foil & Wires	14	.380	13.6	9230	1.5	2.2	3.7	5.2
RG11 Foam Coax, Foil & Braid	14	.405	14.9	9292	1.5	2.2	3.7	5.2

* Not manufactured by Belden. Price based on Saxton type.

The first fact to consider when comparing transmission lines is that foam coax has *lower* losses than any kind of unshielded twinlead when low loss is needed the most: under adverse weather and environmental conditions. Tables 5-1 and 5-2 show that although the theoretical losses for dry, new, perfectly installed standard twinlead are about half those of foam-dielectric coax with No. 20 center conductor, the twinlead losses become many times higher than *any* kind of coax when the line is wet. Oval foam-encased twinlead (Belden 8285) offers greatly improved wet-weather performance, but it is more expensive and difficult to install than low-cost foam coax with comparable wet-weather losses (such as Belden 9275). If an RG6-sized coax (Belden 9283) is considered, cable losses lower than the *best* of the unshielded twinleads in wet weather are available, with the added convenience of noncritical installation! The base price of RG6-sized foam cable is only slightly more than that of quality foam twinlead (Belden 8275 or 8285) even when the cost of accessories (baluns and twinlead insulators) is considered.

For long line runs, for any installation with a rotatable antenna, or for installations involving signal-distribution systems, coaxial cable is cheapest and best. RG59-sized foam coax is ideal for vhf-only installations in all but very long runs in weak-signal areas. The slightly more expensive RG6U foam coax is necessary for vhf-uhf or uhf-only installations in medium- or weak-signal areas. Avoid the expensive and less manageable RG11-sized coax unless important uhf signals are extremely weak and you do not wish to use an antenna preamplifier (for which selection criteria are given in Chapters 6 and 8).

Shielded twinlead also has the advantages of low wet losses and noncritical installation, but at a very high cost, both in terms of money and flexibility. As Table 5-2 shows, Belden 8290 costs nearly twice as much as the coax with the same losses (Belden 9283), and has about twice the cross-sectional area with stiffness to match. Belden 9090 is a much smaller and more flexible style of shielded twinlead, but its losses are far higher than any RG59- or RG6-sized foam coax, and it is much more costly. Shielded twinleads become practical and cost effective only for connecting a nonrotated antenna directly to a single tv set when the line length is short. For short line runs, the relatively high cost of this line is reduced by the cost savings on the two baluns not needed for an all 300-ohm system. Low-loss Belden 8290 is recommended for such installation in weak-signal areas. This line becomes cost effective for lengths of 50 feet or less, but runs of 75 feet can be considered on the basis of convenience alone. If price is no consideration, 8290 is superior to RG6-sized coax at any line length for the reception of very weak uhf signals, since the absence of 1–2 dB balun losses gives 8290 the edge. Belden 9090 is ideal for use in strong-signal areas, where its relatively high losses are of no importance. This line becomes cost effective at lengths of 75 feet or under, but 100-foot line runs can be considered for the convenience alone when using prefabricated lines. Base the line selection on the *weakest* uhf signal you wish to receive. If uhf reception is unimportant, use the highest vhf high-band station.

CHAPTER 6

Signal-Distribution Components

Most of the devices discussed in this chapter either distribute signal to tv receivers served by the system or process the signal in some way before distribution. A few are essentially mechanical devices which simply carry signal. All but the amplifiers are passive devices, i.e., they do not increase the signal power. As such, they must cause an absolute minimum of signal loss when performing their particular function, besides performing that function well. The author tested and measured the performance of many different models of the various types of signal-distribution components to gather valid comparative data. The models specified in this chapter for each job are those that yielded the best performance at the best price in these tests. Bear in mind, though, that these measurements are based on a small sampling, and unit-to-unit variations do exist. Also, not every model component available was tested.

The types of components covered in this chapter are those that are most important to a high-quality, small matv or home system. There are many more types of signal-distribution components in use than are covered here. These other devices are primarily for catv and large matv systems. They were omitted from this book by virtue of being too expensive, too difficult to install or use, or too complex for small matv and home systems. Some omitted devices are at the other extreme: too poor in performance or too amateurish for a quality installation.

The majority of the components discussed in these chapters are 75-ohm devices, or devices that interface with 75-ohm equipment.

This is because nearly all signal-distribution systems use 75-ohm coaxial cable, and 75-ohm devices generally yield the highest performance.

Quality indicators common to most passive components are *insertion loss* and *standing-wave ratio,* or *swr.* Insertion loss indicates how much signal is lost in passing through a passive device; swr indicates how close the actual impedance is to the nominal impedance. In each case, the lower the number, the better the device.

BALUNS

A balun is a broadband transformer with a turns ratio designed to match a 300-ohm system to a 75-ohm system, and vice versa. This device also performs the balanced to unbalanced conversion (and the reverse) customarily required by these systems. Baluns are low cost, and sold by every manufacturer in the tv business!

The most common applications of the balun are to match a 300-ohm antenna to coaxial cable, and to match coaxial cable to the 300-ohm input of a tv set. In this latter application, the main requirement is electrical performance, with physical characteristics secondary. Three of the best indoor baluns are shown in Fig. 6-1. As Table 6-1 shows, the rather large Sony EAC-20W and the RMS Electronics CA-2500 are superb vhf valuns, having measured insertion losses of 0.4 dB over the entire vhf tv band. In fact, the Sony balun will work well to about 700 MHz (Channel 52), above which its swr and losses increase sharply. The AVA Electronics MT-1000B and Saxton S-296 are the same balun (shown at the bottom of Fig.

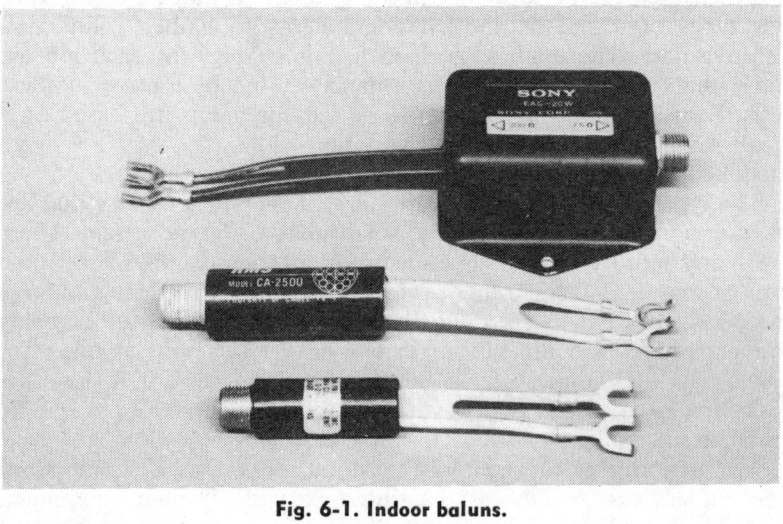

Fig. 6-1. Indoor baluns.

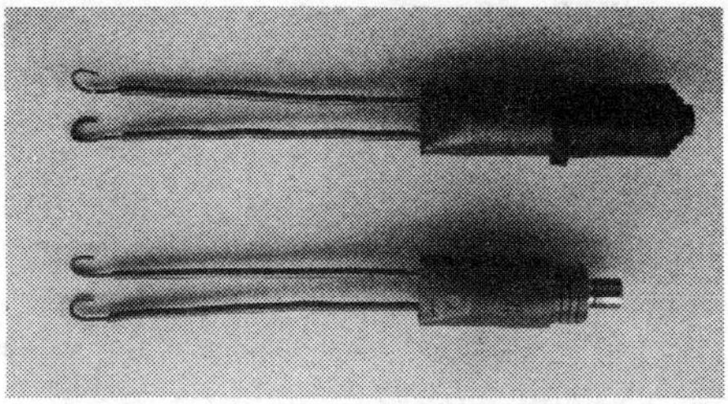

Fig. 6-2. Outdoor "universal" baluns.

6-1) imported by different companies. (Most signal-distribution components are manufactured or assembled in other countries.) They have higher vhf losses than the first two, but their insertion loss is flat to the very end of the uhf band (890 MHz). This, plus low swr, makes them excellent uhf baluns. Their miniature size can also be an asset if the back of the tv is in plain view.

To match a 300-ohm antenna to coaxial cable, the balun requirements are more severe. Not only must the balun have low insertion loss and swr at all frequencies, it must also be waterproof and physically compatible with the terminals of the antenna to which it is connected. Some antennas have widely spaced terminals or an obstruction near the terminals. For instance, the typical 4-bay uhf antenna has a 1¼-inch-diameter tube supporting the bowtie elements. This tube prevents most baluns from being connected from behind the terminals. The only type of balun that can be mounted in this

Courtesy Sony Corp. of America

Fig. 6-3. Lead dress and attachment of ATR-375 balun.

advantageous position is one having long stiff leads, such as the RMS Electronics ATR-375 (Fig. 6-2) or Lafayette 40347. As shown in Fig. 6-3, the leads can be bent to clear the tube, and hold the balun in a position that allows the cable to be tied to the reflector screen of the antenna. This makes a rigid arrangement that places no strain on the balun or antenna terminals. Both of these baluns are encapsulated in plastic and supplied with a rubber boot to weatherproof the cable connection. (Note: the rubber boot on the Winegard CX-series cables also fits these baluns.)

The outdoor balun with the best measured electrical performance is the Jerrold Electronics STO-82, shown in Fig. 6-4. This balun has rigid input terminals, however, so it is physically compatible only with Jerrold antennas. It should always be selected for use with Jerrold antennas, not only for its superb performance, but also for the fact that the mounting clip provides strain relief. It is always advisable to relieve the stress on the balun connection in one way or another. Some outdoor baluns are equipped with mounting clips that fit the popular 1-inch square boom. The Gavin/Antennacraft T-5K balun kit (Fig. 6-5) has a plastic mounting clip and inserts which allow the balun body to be secured to three kinds of antenna boom: 1-inch square, ¾-inch square, or 1-inch round.

SIGNAL SPLITTERS

A signal splitter is a device that permits one antenna (or other signal source) to drive two or more tv sets (Fig. 6-6A) while main-

Fig. 6-4. Jerrold STO-82 balun.

Table 6-1. Measured Performance of Recommended Baluns

Model	Frequency Range	Insertion Loss (dB) 50	200	550	Standing-Wave Ratio 50	200	550	800 (MHz)
Indoor								
RMS CA-2500	VHF	0.4	0.4	—	1.17	1.3	—	—
Sony EAC-20W	VHF	0.35	0.4	0.8	1.08	1.2	1.22	—
AVA MT-1000B	VHF-UHF	0.6	0.7	0.8	1.17	1.3	1.7	1.55
Outdoor								
RMS ATR-375	VHF-UHF	0.45	0.4	0.9	1.17	1.31	1.33	1.8
Laf. Radio 40347	VHF-UHF	0.45	0.4	0.9	1.17	1.31	1.33	1.8
Antennacraft T-5	VHF-UHF	0.5	0.7	1.0	1.16	1.5	1.6	1.6
Special								
Jerrold STO-82	VHF-UHF	0.65	0.55	0.5	1.3	1.2	1.35	1.3

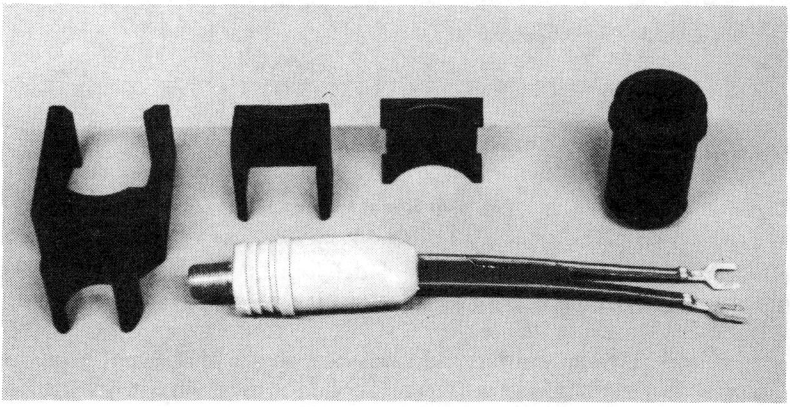

Fig. 6-5. Gavin/Antennacraft T-5K balun kit.

taining the proper impedance match at all ports and isolating each tv set connected to it from the other(s). Resistive devices that perform this task cause a power loss, so are not used in quality systems. High-grade tv signal-distribution systems use the hybrid splitter, a device that accomplishes power division and impedance matching with negligible total power loss.

The prime quality ratings of a hybrid splitter are insertion loss, output-to-output isolation, and swr. The insertion loss to each output of a splitter is the sum of the theoretical splitting loss and the actual power lost in the splitter. The theoretical signal loss to each output depends on the number of outputs; it is 3 dB for a two-way splitter, 4.8 dB for a three-way splitter, 6 dB for a four-way splitter. The real power loss of the splitter will result in the total power loss to each output being always somewhat greater than the theoretical losses just mentioned. For instance, a perfect two-way splitter will have 3.0 dB loss to each output, an excellent splitter might have 3.1 dB

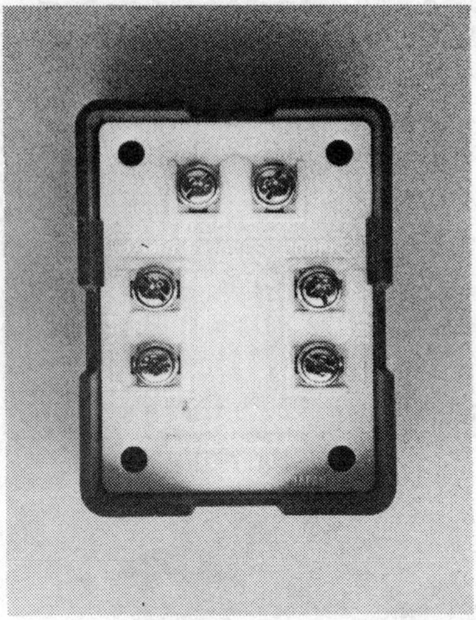

Courtesy Sony Corp. of America

(A) Splitter application.

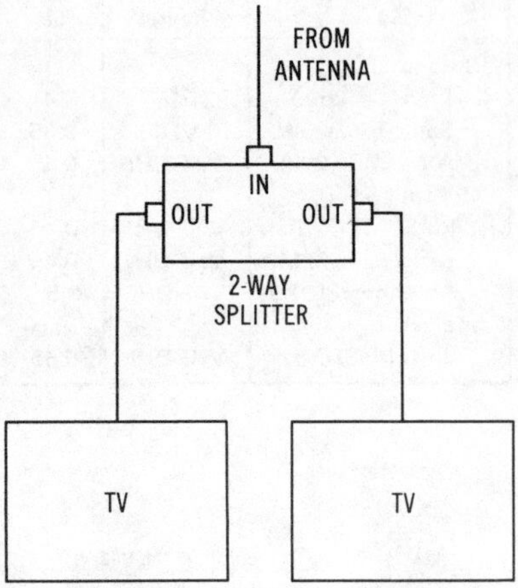

(B) 300-ohm 2-way splitter.

Fig. 6-6. Signal splitter.

loss, and a poor splitter possibly 4 dB loss at vhf frequencies. The losses at uhf frequencies are always higher, even for high-quality splitters, as shown in Table 6-2.

Isolation between outputs indicates how much of a signal (such as the oscillator radiation of a tv set) applied at one output will appear at the other output(s). The higher the isolation, the better the splitter. As Table 6-2 shows, many of the recommended splitters have very high isolation figures. The "hitch" in this idyllic situation is that each output port must be perfectly terminated to achieve the high isolation figures and low swr. Unfortunately, the terminating impedance provided by the input of a tv set is far from perfect; it depends greatly on which channel the set is tuned to, whether or not the power is turned on, and the quality of the engineering job. So, while splitters do an excellent job in laboratory tests, the signal-wasting directional coupler (covered later in this chapter) works better in real-life applications. The signal splitter is, however, used in large signal-distribution systems to feed multiple branch lines of directional couplers. And signal "splitters" are also used as signal combiners, as you will soon see.

300-Ohm Splitters

Splitters are available in either 75-ohm or 300-ohm impedances. The 300-ohm splitters cost less, but have commensurately lower

Table 6-2. Measured Performance of Signal Splitters

Outputs	Impedance (Ω)	Model	Insertion Loss (dB) 50	200	550	800	dB Isolation Between Outputs 50	200	550	Input Standing-Wave Ratio 50	200	550	800 (MHz)
2	300	RMS C-2UV	3.1	3.1	3.7	—	16	19	21	—	—	—	—
	75	RMS MA-2UV	3.1	3.1	3.4	3.8	21	34	31	1.07	1.15	1.07	1.08
		RMS CA-1002/SM	3.0	3.0	—	—	33	38	—	1.06	1.16	—	—
3	75	RMS MA-30U/SM	5.2	5.2	5.1	5.6	24	31	40	1.09	1.11	1.23	1.22
		RMS CA-2003/SM	4.9	5.1	—	—	32	31	—	1.00	1.20	—	—
4	75	Jerrold 1597A	6.4	6.4	6.6	7.5	42	38	40	1.32	1.36	1.04	1.35
		RMS MA-4UV	6.8	6.7	6.8	7.0	27	34	34	1.12	1.10	1.03	1.25
		RMS CA-1004/SM	6.6	6.8	—	—	38	30	—	1.10	1.12	—	—
		RMS CA-2004/SM	6.3	6.4	—	—	35	28	—	1.06	1.10	—	—

performance. For noncritical applications where the tv receivers are fairly close together, a two-way 300-ohm splitter can be used. The RMS Electronics C-2UV (Fig. 6-6B) is the all-channel two-way splitter with the lowest measured insertion loss and highest isolation. The C-2UV can be used outdoors if it is mounted with the terminals facing downward. Four-way 300-ohm splitters are also available.

75-Ohm Splitters

Seventy-five-ohm splitters offer far better performance than do 300-ohm units. Table 6-2 shows that insertion loss at vhf frequencies is close to theoretical in many cases, and isolation and swr figures of the vhf-uhf models are quite good even at uhf frequencies.

Splitters of different numbers of outputs are available for vhf only, or all-channel operation. All-channel (vhf-uhf) splitters can be used in place of vhf-only splitters for vhf applications. However, all-channel splitters are more expensive than their vhf-only counterparts, and their vhf performance (particularly low-end isolation) is often inferior to that of the vhf-only splitter. At the tops of Figs. 6-7A and 6-7B vhf-only splitters are shown, while their all-channel counterparts are shown at the bottoms of these photographs. All are waterproof units from RMS Electronics. A point to remember is that three-way splitters are somewhat rare, especially when they provide three equal and minimum-loss outputs.

Four-output splitters are available in many configurations, as is evident from Fig. 6-7C. The left-most splitter is the RMS CA-2004/SM vhf-only splitter. The center and right-most splitters are the RMS MA-4UV and Jerrold 1597A all-channel splitters. The RMS MA-4UV and its vhf equivalent, the CA-1004/SM, combine low cost and good performance. Like all of the RMS Electronics splitters shown in this chapter, they feature a cast one-piece zinc case, so they can be used outdoors if desired. The Jerrold 1597A is much more expensive, but provides top performance at all frequencies. Note that its vhf isolation figures in Table 6-2 are even better than those of the top vhf-only splitter. The 1597A requires a weatherproof housing for outdoor use.

Splitters with eight outlets are also available for matv applications.

Signal Combiners

Hybrid signal splitters can be used "backwards" to combine the signal from two or more identical antennas to produce a stronger signal. For this to be worthwhile, the splitter must have very little internal power loss at the frequencies of interest. Also, the splitter should be small sized and waterproof, since these devices will be located outdoors near the antennas. Table 6-3 shows the measured combining gain at several frequencies for splitters well suited to this

(A) Two-way splitters.

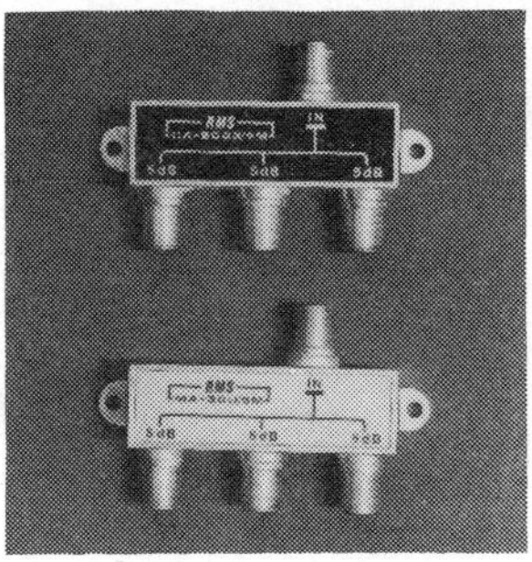

(B) Three-way splitters.

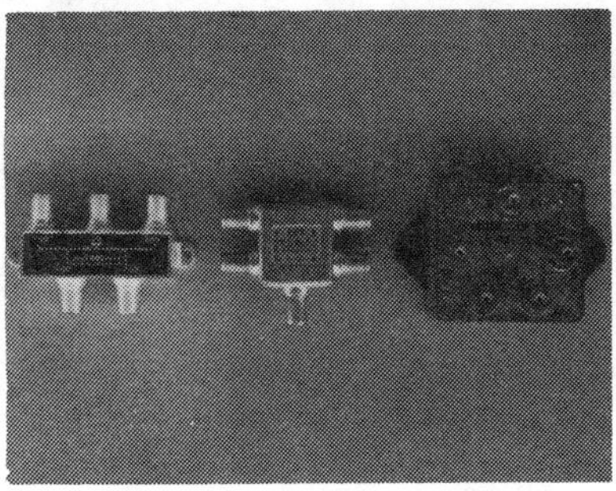

(C) Four-way splitters.

Fig. 6-7. Two-, three-, and four-way 75-ohm signal splitters.

purpose. In Chapter 8 you will see how to apply these devices for a stronger signal.

BAND SEPARATORS

A band separator is a device that separates the uhf and vhf signals coming from an all-channel antenna, and then applies the separated

Table 6-3. Measured Combining Gain of Signal Splitters

Inputs	Impedance (Ω)	Model	Combining Gain (dB) 50	200	550	800 (MHz)
2	300	RS C-2UV	2.2	2.2	—	—
	75	RMS MA-2UV	2.7	2.5	2.5	2.2
		RMS CA-1002/SM	2.8	2.6	—	—
4	75	RMS MA-4UV	5.2	5.3	5.4	5.0
		RMS CA-1004/SM	5.5	5.5	—	—
		RMS CA-2004/SM	5.4	5.4	—	—

signals to the proper input terminals of the tv set (Fig. 6-8). This action eliminates the inconvenience of manually switching the transmission line from one set of input terminals to the other when changing between vhf and uhf channels.

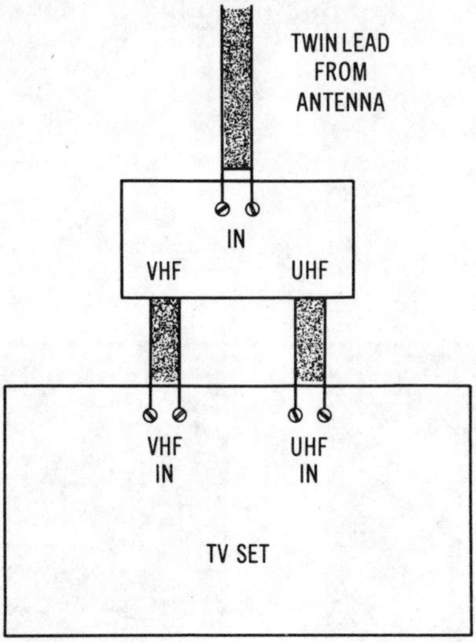

Fig. 6-8. Application of a band separator.

Band separators are available in several impedance combinations, so a suitable band separator is available for nearly every application. Most band separators are designed to attach directly to the tv set, so both the vhf and uhf outputs are usually 300 ohms impedance.

The all-300-ohm band separator shown in Fig. 6-9 is ideal for a small tv system (one or two sets) where the transmission line is twinlead. The Channel Master 0012 costs slightly more than the usual 300-ohm-input model, but its performance is so superior to the rest that it is a bargain at twice the price. The 0012 has 10–15 dB more out-of-band rejection than any of the other 300-ohm models

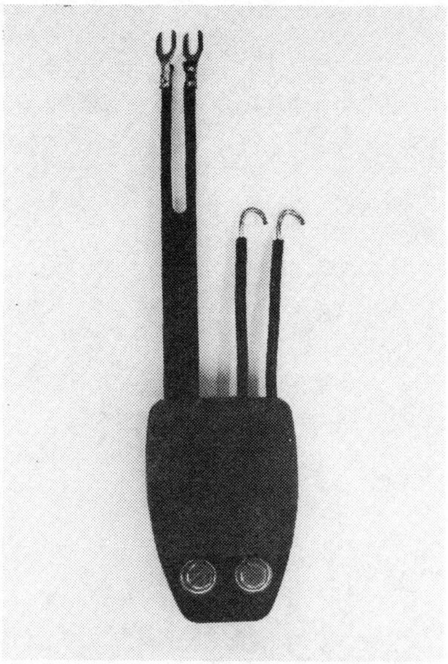

Fig. 6-9. 300-ohm vhf-uhf band separator.

tested. In fact, as Table 6-4 shows, its performance is comparable to the 75-ohm band separators.

The Saxton S-314 and RCA 10G221 (Fig. 6-10) and the Arista 900 have a 75-ohm input for use with coaxial cable. The excellent performance of the specified models is due to their having 75-ohm

Table 6-4. Measured Performance of VHF-UHF Band Separators

Model	Input/Output Impedance (Ω)	Port	Insertion Loss (dB) 100	200	470	550 (MHz)
Ch. Master 0012	300	V	0.2	0.2	32	35
		U	36	27	−0.1	0.7
RCA 10G223	300/75	V	0.5	0.7	26	28
		U	37	25	0.7	0.8
RCA 10G221	75/300	V	0.5	0.6	28	33
		U	35	30	0.7	1.2
Arista 900	75/300	V	0.5	0.6	28	33
		U	35	30	0.7	1.2
Saxton S-314	75/300	V	0.4	0.4	20	35
		U	40	24	1.0	1.2
Winegard CS-775	75	V	0.0	0.2	49	49
		U	53	36	0.6	0.0
Finco G-520	75	V	0.0	0.5	35	30
		U	38	20	0.4	0.3
Finco G-522	75	V	0.1	0.5	25	31
		U	37	18	0.6	0.4

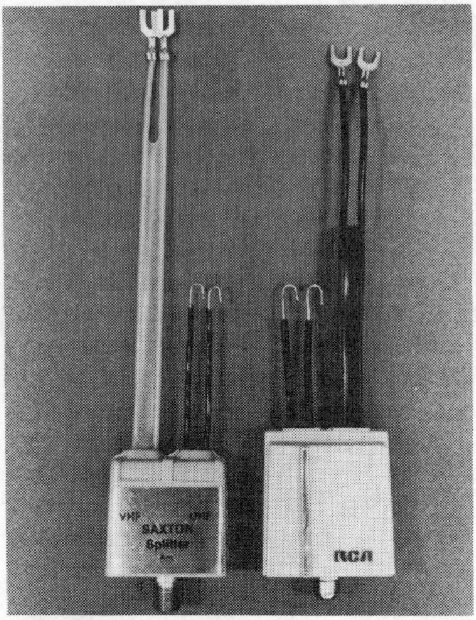

Fig. 6-10. 75-ohm input vhf-uhf band separator.

filter circuitry and baluns in each output lead to provide the 300-ohm outputs. Most 75-ohm input band separators use simple 300-ohm circuitry and a single balun at the input, an arrangement that results in the relatively poor performance of the *usual* all 300-ohm device.

The RCA 10G223 (Fig. 6-11) is a band separator identical internally with the 10G221, but designed for outdoor use. Thus it can

Fig. 6-11. RCA 10G223 band separator/antenna combiner.

be used "backwards" as a *combiner* for separate uhf and vhf antennas, enabling them to share one coaxial downlead to the distribution system. In this application the uhf and vhf "output" ports become antenna *inputs,* and the common coaxial downlead is connected to the all-channel "input" port.

For the ultimate in band separation and the lowest insertion loss, an all-75-ohm band separator is unbeatable. However, a balun is required to match at least the uhf outlet to the tv set, since very few tv sets built for the U.S. market have a 75-ohm uhf input. The all-75-ohm band separator is most often used in complex matv systems, and as an antenna combiner. The Winegard CS-775 (Fig. 6-12) is a low-cost unit that provides nearly loss-free in-band transmission and extremely high out-of-band rejection (in the neighborhood of 50 dB!). The Finco G-520 (Fig. 6-12) is another low-cost unit with good performance. Its ports all face in the same direction, so it will fit in places where the CS-775 will not, and vice versa.

Fig. 6-12. 75-ohm vhf-uhf band separator/combiners.

For outdoor use (to combine uhf and vhf antennas), the very convenient Finco G-522 (Fig. 6-13) can be used instead. The G-522 is specifically designed for combining antennas, and is equipped with a universal mounting bracket and fittings that permit easy attachment of this weatherproof unit to the antenna mast or boom. Rubber boots to weatherproof the cable connections are also supplied.

The Winegard F-175 (Fig. 6-14) is a low-cost, all-75-ohm band separator intended for separating the vhf high and low bands. For outdoor use the choice of vhf H/L band separator/combiner depends on the antenna arrangement. If the high- and low-band antennas are

Fig. 6-13. Mast-mounted 75-ohm band separator/combiner.

mounted on the same mast, the Finco G-514 is best. This unit has the same high-quality, cast-metal weatherproof enclosure as the G-522, so it can be mounted right on the mast near the antennas. If the antennas are mounted on separate masts (as large arrays often are), the Winegard F-175 in a WH-1 weatherproof enclosure (Fig. 6-15) should be considered. Performance data on both units are given in Table 6-5.

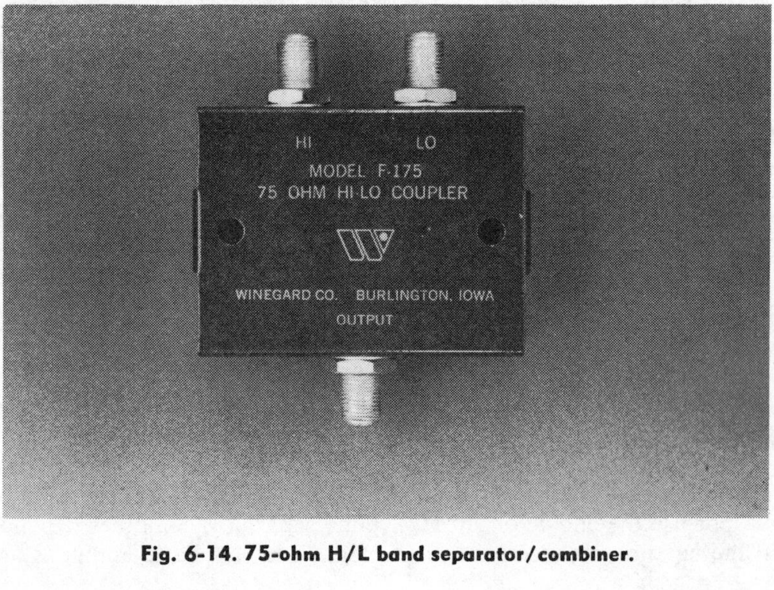

Fig. 6-14. 75-ohm H/L band separator/combiner.

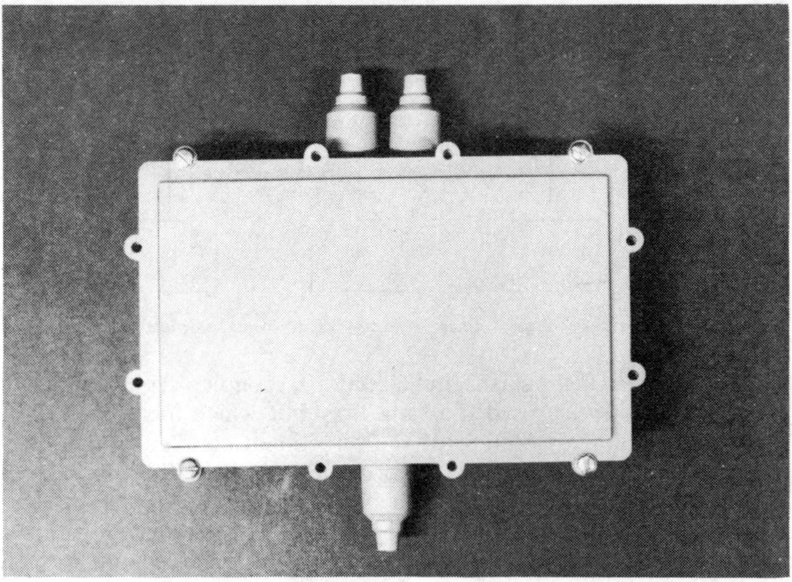

Fig. 6-15. Winegard WH-1 weatherproof enclosure.

DIRECTIONAL COUPLERS

A directional coupler is a device whose signal outlet is well-isolated from the transmission line. Directional couplers are used when many tv sets are fed by a high-level signal. All directional couplers in the system are connected in tandem, as shown in Fig.

Table 6-5. Measured Performance of VHF H/L Band Separators

Model	Impedance (Ω)	Port	Insertion Loss (dB) 50	90	110	175	200 (MHz)
Winegard 175	75	L	55	42	30	0.6	0.2
		H	0.1	0.2	0.4	30	36
Finco G-514	75	L	36	25	24	0.3	0.2
		H	0.1	0.3	0.4	24	18

6-16. The collection of cables connecting the directional couplers is known as the *trunk line*. The short cables connecting the directional couplers to the tv sets are called *drop cables*. Unlike a splitter, a mistermination at the end of a drop cable fed by a directional coupler will not affect the signal fed to the other tv receivers. The only requirement is that a 75-ohm termination be connected to the OUT port of the last directional coupler in the chain.

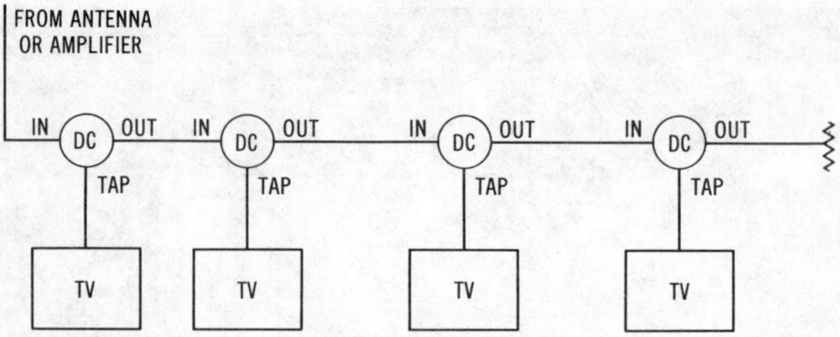

Fig. 6-16. Signal distribution by directional coupler.

Two features distinguish the directional coupler from the *drop tap,* a resistive device used the same way but which has poorer performance. First, the isolation of the directional coupler from the tv port (called the *tap*) to the OUT port (and thus the other directional coupler down the trunk line) is much greater than the signal attenuation from trunk line to tv port (tap). The isolation of a drop tap is the same as its attenuation. Second, a 75-ohm impedance is seen looking into every port on the directional coupler. This condition is known as *back matching.* Drop taps are not back matched.

The directional coupler shown at the top of Fig. 6-17A is one of the RMS Electronics CA-1090/M series. These hybrid back-matched couplers not only provide excellent performance (see Table 6-6), but are low priced and waterproof! Although rated to only 300 MHz, the author's measurements show the higher-value directional couplers provide adequate performance over the lower part of the uhf band. The CA-1090/M series is available in 6, 9, 12, 16, 20, 24, and 30 dB tap-attenuation values.

(*A*) *Single-tap types.*

(*B*) *Wall-tap style.*

Fig. 6-17. Directional couplers.

Blonder-Tongue's TO-1 series directional couplers (middle of Fig. 6-17A) provide very good performance over the vhf band, despite a very low price tag. These waterproof directional couplers are available in 12, 16, 20, and 24 dB tap values.

For superb performance over the vhf and *full* uhf bands, the Jerrold DC-12B and DC-16B directional couplers (bottom of Fig. 6-17A) are unbeatable though relatively expensive. Jerrold also has a modestly priced series of flush-mount vhf-uhf wall taps (actually directional couplers) that is a little more expensive. The DFT series (Fig. 6-17B) is available in 13, 19, 25, and 31 dB values of tap attenuation. Though not as high in performance as the directional coupler, the DFTs perform extremely well through the entire uhf band. These directional couplers are intended to mount behind wall plates, either in-wall or in outlet boxes (Fig. 6-17B).

Directional couplers with multiple taps (tv ports) are also available. These money-saving devices are useful in large systems where drops to adjacent rooms can be made at their common corner, or where several tv sets are located fairly close together. As was the case with the single-tap directional couplers, the RMS Electronics CA-2014T series of four-tap directional couplers (top of Fig. 6-18A) is recommended for vhf installations, and the Jerrold DCT4 series (bottom of Fig. 6-18A) is recommended for all-channel installations. The choice of two-tap directional couplers is much more restricted; the RMS Electronics CA-2012 series (Fig. 6-18B) is one of the few available. The performance of the samples of these devices tested by the author is included in Table 6-6.

AMPLIFIERS

Amplifiers are *active* devices: They increase the power level of signals passing through them. Amplifiers used in home tv and small matv systems usually fall into either of two categories—preamplifiers and distribution amplifiers. Preamplifiers are intended to increase the level of a weak signal (20–200 microvolts per channel) before the signal is applied to a high-loss component (transmission line) or a high-noise device (the uhf input on a tv set). Distribution amplifiers are driven by much higher-level signals (500–20,000 microvolts per channel); the idea here is to boost this to an even higher level (10,000–100,000 microvolts per channel) to drive a high-level signal-distribution system (like that shown in Fig. 6-16).

The most important performance criterion for a preamplifier is its *noise figure,* a method of indicating how much (or how little) noise the amplifier adds to the signal before amplifying it. The smaller the noise figure, the better the preamplifier. High gain is useless unless the amplifier noise is low. If a weak signal with decent signal-

(*A*) *4-tap types.*

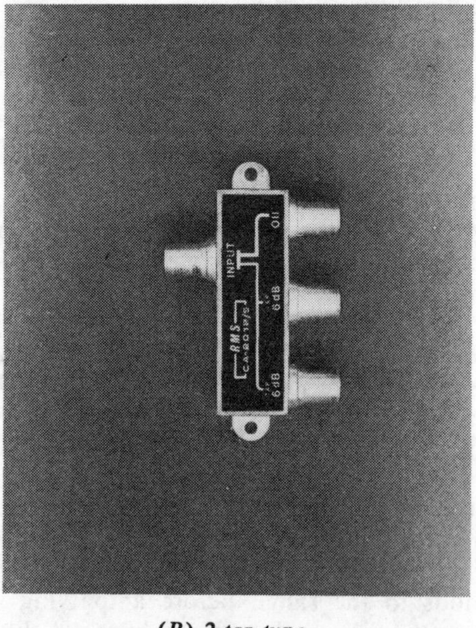

(*B*) *2-tap type.*

Fig. 6-18. Multiple-tap directional couplers.

Table. 6-6. Measured Performance of Recommended Directional Couplers

Taps	Model	Insertion Loss (dB)			Tap Loss (dB)				Tap Isolation (dB)				Input SWR				Tap SWR			
		50	200	550	50	200	550	800	50	200	550	800	50	200	550	800	50	200	550	800
1	RMS CA-1090- 6	1.6	1.5	—	7	7.1	—	—	31	30	—	—	1.17	1.14	—	—	1.14	1.21	—	—
	RMS CA-1090-12	1.0	0.8	1.0	12	11.7	12.5	—	31	31	25	—	1.20	1.17	1.03	—	1.22	1.22	1.56	—
	RMS CA-1090-16	0.4	0.4	0.6	16	15	17	—	40	32	24	—	1.03	1.08	1.20	—	1.22	1.30	1.55	—
	B-T TO-1-12	0.8	0.7	0.6	12	12.2	12	—	59	34	22	—	1.12	1.12	1.28	—	1.08	1.06	1.19	—
	Jerrold DC-16B	0.8	0.5	0.8	16	16	15	15	47	36	38	30	1.07	1.13	1.17	1.10	1.11	1.09	1.12	1.25
	Jerrold DTF-19	0.2	0.3	0.1	19	19	18.5	19	48	35	26	23	1.03	1.08	1.25	1.17	1.02	1.06	1.08	1.12
2	RMS CA-2012-6	3.2	3.2	—	6.8	6.7	—	—	33	36	—	—	1.06	1.18	—	—	1.12	1.08	—	—
4	RMS CA-2014T-15	1.0	1.0	1.1	15	15	18	—	39	38	36	—	1.10	1.12	1.23	—	1.06	1.07	1.35	—
	Jerrold DCT4-14	1.4	1.4	2.1	14	14	14	15	40	30	23	20	1.14	1.07	1.25	1.50	1.22	1.27	1.22	1.45

to-noise ratio is applied to a noisy preamplifier, the preamplifier noise is boosted along with the signal, and the result is a high-level signal with high-level noise (poor signal-to-noise ratio). This appears on the tv screen as a strong (high-contrast) signal with lots of snow!

The most important criterion for a distribution amplifier is output level. If the antenna system is properly designed, the signal level to the distribution amplifier will be high enough (1000–5000 microvolts) that its own noise will have little effect on the signal-to-noise ratio of the output signal. Instead, the distribution amplifier must be able to put out enough signal (possibly 100 millivolts per channel) to drive a high-level distribution system.

A selection criterion important to both types of amplifier is swr. The input and output impedances of a distribution amplifier must stay as close to 75 ohms as possible over the entire passband of the amplifier to preclude the generation of line reflections (ghosts). Similarly, the input impedance of a mast-mounted preamplifier must closely approximate 300 ohms to provide a good match to the antenna for maximum power transfer. Low swr is a lot harder to achieve in amplifiers than in some of the passive devices; an swr of 1.2–1.5 is considered very good for an amplifier.

Mast-Mounted Preamplifiers

The signal-to-noise ratio is higher at the antenna terminals than at any other place in the system. Furthermore, the signal level is higher at the antenna terminals than it is at the output end of the transmission line. Since the input to a preamplifier should be as high as possible to minimize the effect of the preamplifier's own noise, mounting the preamplifier right at the antenna enables it to put out a signal with a much better signal-to-noise ratio than if it were mounted at the output end of the transmission line.

The main value of a preamplifier lies in its ability to boost weak-to-marginal signals before they are further weakened by transmission-line losses. Another benefit is the improvement in picture signal-to-noise ratio due to the decibel difference between the tv tuner's noise figure and that of a modern low-noise preamplifier. For uhf reception this effect is especially dramatic. The noise figure of the typical uhf tuner is about 13–14 dB, so a very high signal level is needed to produce a noise-free picture. A very-low-noise preamplifier can boost the signal high enough to almost eliminate the effect of uhf tuner noise without adding enough noise of its own to nullify this achievement.

Preamplifiers should not be used in strong-signal areas. Unlike passive components, preamplifiers are subject to overload. Overload of a preamplifier (or distribution amplifier) depends not only on the

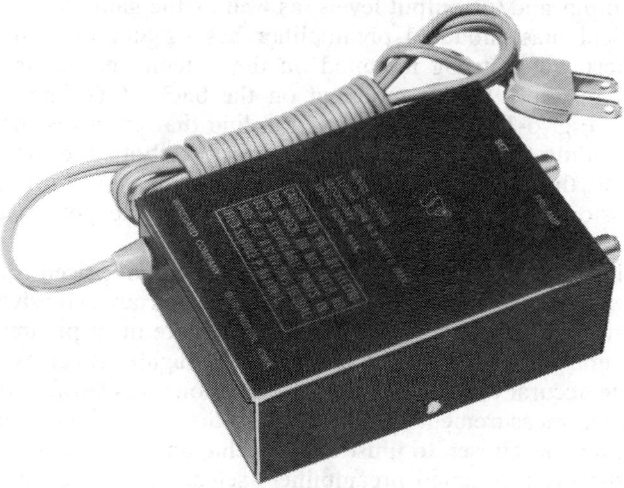

(*A*) *Power supply.*

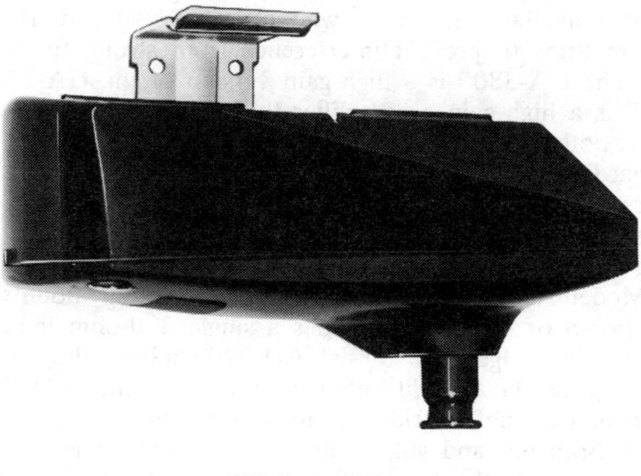

(*B*) *Preamplifier.*

Courtesy Winegard Co.

Fig. 6-19. Typical mast-mounted preamplifier and power supply.

strength of the tv signals accepted by the preamplifier's bandpass circuits, but also on the number of these tv signals. The sum of the individual signal levels supplied to the preamplifier must not exceed the specified input level (or the output level divided by the preamplifier gain). Reputable preamplifier manufacturers always specify the maximum input and/or output levels, as well as the gain.

The typical mast-mounted preamplifier has its amplifier circuitry in a weatherproof housing mounted on the antenna boom or mast, and a power supply that is mounted on the back of (or near) the tv receiver (Fig. 6-19). The transmission line that connects the preamplifier module to the power supply transports both the amplified tv signal and the power to operate the preamplifier. A short length of transmission line then conveys the signal from the power-supply module to the tv receiver.

The Winegard Company makes several medium-priced quality preamplifiers like the one shown in Fig. 6-19, and an extensive line of high-performance cartridge preamplifiers. Unlike most preamplifier manufacturers, the Winegard Company thoroughly specifies their devices. The accuracy of the claimed specifications was verified by the author's own measurements on a few of their amplifiers, so the Winegard line was chosen to illustrate this chapter. The specifications for all of the mast-mounted preamplifiers selected for discussion are listed in Table 6-7. All of these preamplifiers have 300 ohms input impedance.

The GA-3700 provides medium gain on the vhf channels, and passes uhf with some loss. This model is for coaxial downlead; the GA-3000 is available for those who insist on twinlead. Both units contain fm traps to prevent interference from strong fm broadcast stations. The GA-3800 is a high-gain version of the GA-3700. The GA-8800 is a high-gain preamplifier that provides low-noise amplification of both vhf and uhf frequencies.

Winegard's AC series is a line of high-performance cartridge preamplifiers (Fig. 6-20A) that plug into housings in certain of their antennas. These preamplifiers can be used with any model antenna by using one of the universal housings shown in Fig. 6-20B.

The Model AH-0100 universal cartridge housing mounts on the antenna boom or mast and provides a single 300-ohm input to the preamplifier cartridge. The Model AH-0200 allows the installer to couple separate uhf and vhf antennas directly to the individual preamplifiers of a vhf-uhf cartridge without the use of an external antenna combiner. Both uhf and vhf inputs are 300 ohms. The Model AH-0300 is the same as the AH-0200, but both inputs are 75 ohms.

The AC-4990 is a uhf-only preamplifier in a class by itself. No other uhf preamplifier manufactured today (1977) has a comparable noise figure. The incredibly low 2.2-dB noise figure of this medium-gain preamplifier permits a dramatic improvement in poor- and marginal-signal areas over antenna systems not using any preamplifier. A poor picture (in regard to snow) will become good; a fair picture will become excellent. The 4990 passes vhf frequencies with slight

Table 6-7. Specifications for Selected Winegard Mast-Mounted Preamplifiers

Characteristic		UHF AC-4990	VHF-UHF AC-9880	VHF-UHF AC-9990	VHF-UHF GA-8800	VHF GA-3800	VHF GA-3000	VHF GA-3700	VHF AC-2950	VHF AC-9130	VHF-H AC-9730	VHF-L AC-9260
Bandpass (MHz)	VHF	54–216	54–216	54–216	54–216	54–216	54–216	54–216	54–216	54–216	174–216	54–88
	UHF	470–890	470–890	470–890	470–890	470–890	470–890	470–890	470–890	470–890	—	—
Gain (dB)	VHF	−0.8	14	25	26	26	15	15	14	24	24	26
	UHF	17.5	17.5	17.5	18	−2.5	−4	−4	−1.6	−1.6	—	—
Noise Figure	VHF	—	3	3	3.4	3.4	3	3	3	3	3	2.2
	UHF	2.2	2.2	3	3.5	—	—	—	—	—	—	—
Max Total	VHF	—	1000	1100	800	800	1000	1000	1000	1200	1200	1200
Output (mV)	UHF	900	900	980	630	—	—	—	—	—	—	—
Max Total	VHF	—	200	75	40	40	180	180	200	80	80	80
Input (mV)	UHF	130	130	130	80	—	—	—	—	—	—	—
Output Impedance (Ω)		75	75	75	75	75	300	75	75	75	75	75
FM Trap		No	Sw	Sw	Fixed	Fixed	Fixed	Fixed	Sw	Sw	No	Sw
List Price ($)		78	91	94	81	57	41	41	47	57	56	56

Sw = Switchable

loss, so this preamplifier is ideal for use with all-channel antennas where vhf reception is satisfactory, but uhf reception is poor (a common circumstance). In combination with a high-gain uhf antenna, such as a 4-bay or the Antennacraft G-1483, it makes a potent long-distance installation. The companion amplifier, the AC-

(A) Winegard AC-series preamplifier.

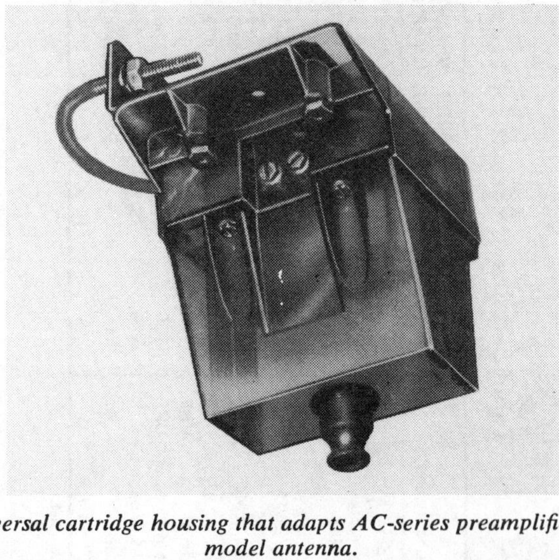

(B) Universal cartridge housing that adapts AC-series preamplifiers to any model antenna.

Courtesy Winegard Co.

Fig. 6-20. Preamplifier cartridge and housing.

9880, has the same uhf circuitry as the 4990, plus a medium-gain vhf preamplifier with a switchable fm trap. The high-quality performance of these units does not come cheap, however; the 9880 plus a universal housing lists at over $100.

The vhf section of the AC-9880 is available separately and at relatively low cost as the AC-2950. This preamplifier provides medium vhf gain, and passes uhf frequencies with a small loss. The AC-9130 is a high-gain vhf preamplifier that also passes uhf frequencies with a small loss. Even though the output capability of the 9130 is high, the total maximum input level is only 80 millivolts because of the high gain. This amplifier is therefore most useful in weak-signal areas. As is the case with the other AC-series vhf amplifiers, the internal fm trap can be switched out of the circuit for full-gain fm reception. This allows the preamplifier to be used with a uhf/vhf/fm antenna in a system that will also drive an fm receiver.

The AC-9730 and AC-9260 are high-gain single-band vhf preamplifiers intended for use with single-band or single-channel antennas. They will not pass uhf frequencies or the other vhf band. The same input-level precaution described for the 9130 should be observed for these preamplifiers.

Indoor Preamplifiers

The noise figure of the typical uhf tuner is so bad that a worthwhile improvement in picture signal-to-noise ratio is possible if an ultralow-noise amplifier is installed between the transmission line and the tv set. Although this is normally the location for a distribution amplifier, the result is that of a preamplifier. The Winegard UA-4050 (Fig. 6-21A) is a device that functions in this location. It has 8½ dB gain and a noise figure of only 3.2 dB at 75 ohms. It performed well in both field and laboratory tests, significantly improving picture quality with very weak to marginal signals. The 4050 contains an excellent band separator, so it can also be used with an all-channel antenna. The antenna cable is connected to the 4050 input, and the tv receiver's 300-ohm vhf and uhf antenna terminals are then connected to the corresponding 4050 output terminals with short lengths of twinlead. Winegard's UA-4030 (Fig. 6-21B) is a similar unit with a 3.7-dB noise figure and 300-ohm input impedance.

Full manufacturers' specifications for both models and other indoor amplifiers are listed in Table 6-8. Notice that the maximum input and output levels for indoor units (particularly distribution amplifiers) are specified on a *per channel* basis. This is more appropriate for distribution amplifiers since in large systems the signal levels are fairly equal by the time they reach the distribution amplifier stage. The levels assume that seven vhf channels and/or five uhf channels

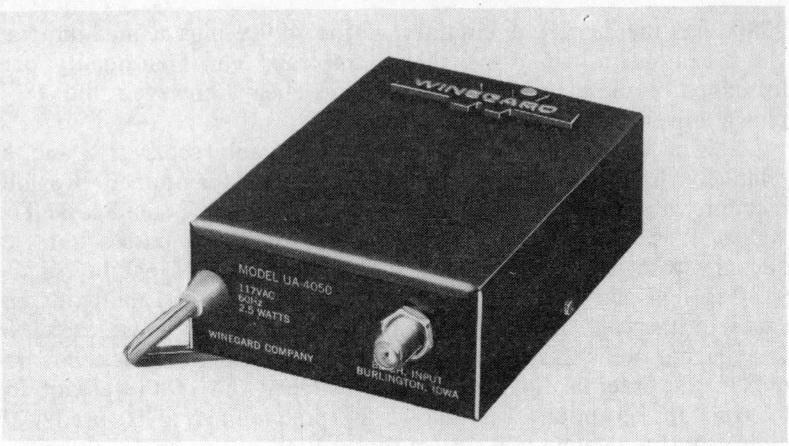

(*A*) *UA-4050.*

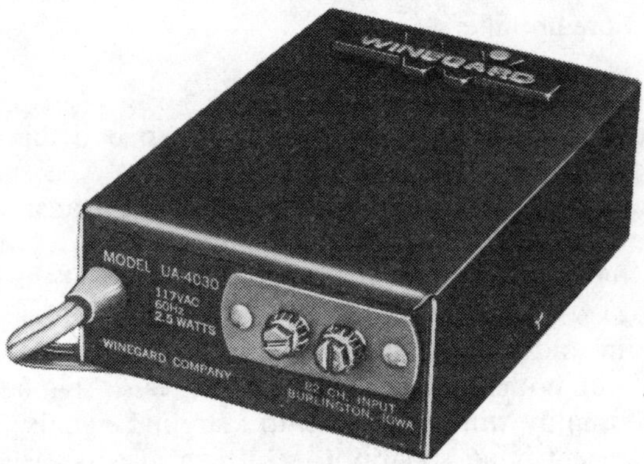

(*B*) *UA-4030.*

Courtesy Winegard Co.

Fig. 6-21. Uhf indoor preamplifiers.

are carried. If the actual number in either band is less, higher signal levels for each channel are permissible. The maximum per-channel signal level for a *lesser number* of channels is easily calculated. First, multiply the rated per-channel signal level (in millivolts) by the stated number of channels (usually seven for vhf, five for uhf). Second, divide the resulting number by the number of channels actually handled by the amplifier. The answer to the second operation is the new per-channel rating in millivolts.

Table 6-8. Specifications for Selected Winegard Indoor Amplifiers

Characteristic		UHF UA-4030	UHF UA-4050	VHF-UHF DA-815	VHF DA-205
Bandpass (MHz)	VHF	54–216	54–216	54–216*	54–300*
	UHF	470–890	470–890	470–890	—
Gain (dB)	VHF	−1.2	−0.6	17	16
	UHF	8	8.5	15	—
Noise Figure	VHF	—	—	2.2	2.9
	UHF	3.7	3.2	6.5	—
Output mV	VHF	—	—	200	250
per Channel	UHF	100	100	200	—
Input mV	VHF	—	—	30	35
per Channel	UHF	35	35	33	—
Input Impedance (Ω)		300	75	75	75
Output Impedance (Ω)		300/300	300/300	75	75
SWR		1.6	1.6	1.4 avg	1.4
FM Trap		No	No	Fixed	Fixed
List Price ($)		51	51	58	33

* Except for portion of fm band.

Distribution Amplifiers

The Winegard DA-815 is an all-channel distribution amplifier suitable for off-the-air signals in areas having both vhf and uhf stations. This low-cost unit provides 15–17 dB gain at the standard channel frequencies, and contains an fm trap. The vhf noise figure of the DA-815 is one of the lowest in the industry for a broadband amplifier. While not in the same league as the UA-4050, the 6.5-dB uhf noise figure is better than necessary for a distribution amplifier, and will even improve the picture signal-to-noise ratio somewhat with weak signals.

(A) DA-205.

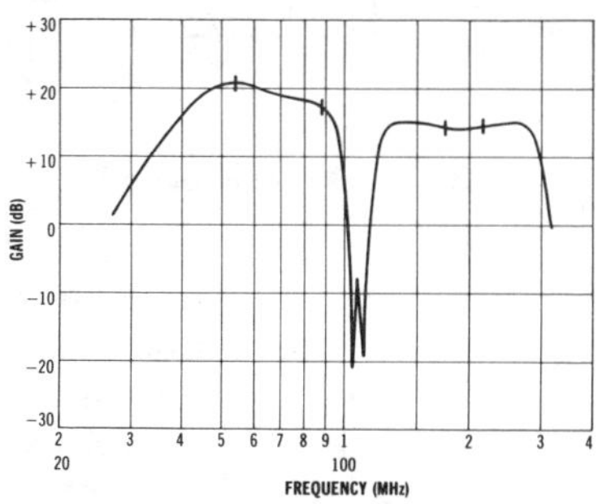

(B) Measured frequency response.

Courtesy Winegard Co.

Fig. 6-22. Winegard DA-205 distribution amplifier and its frequency response.

The Winegard DA-205 (Fig. 6-22A) is a low-noise distribution amplifier covering the vhf and superband channels. This low-cost unit provides 17–21 dB measured low-band gain, and 14–15 dB high-band gain; Fig. 6-22B shows a plot of the measured frequency response. Its noise figure of 3 dB is one of the lowest in the industry for this type amplifier. This gain and noise-figure combination allows a marginal signal (300 microvolts) to drive the distribution system without degrading the signal-to-noise ratio.

ATTENUATORS

Excessive signal, on one or all channels, is a problem in some tv antenna systems. One method of correcting this is with an attenuator of one type or another.

Attenuator Pads

In metropolitan areas excessive signal strength is frequently the problem. Although most tv receivers are capable of withstanding enormous signal levels without ill effect, some are susceptible to cross modulation. Older sets with bipolar-transistor tuners are particularly prone to overload, especially when strong CB or fm signals are also present. The quickest cure for this is to insert an attenuator in the transmission line feeding the tv set or distribution system.

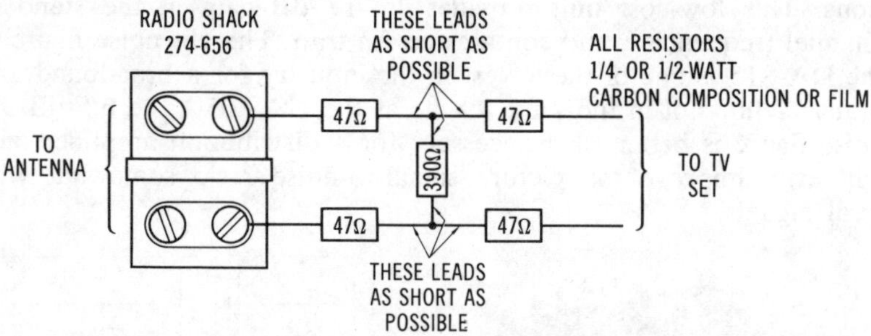

Fig. 6-23. Homemade 6-dB attenuator pad and connecting block.

There are no fixed 300-ohm attenuators readily available; you must make your own. Fig. 6-23 shows how to construct a 6-dB pad from five carbon-composition or carbon-film resistors. (Do not use any other kinds.) A 20-dB pad has the same configuration, but uses 120-ohm resistors instead of 47 ohms, and a 68-ohm resistor instead of 390 ohms. The resistor leads going to the tv set can be left full length, but those at the resistor junctions must be as short as possible and soldered.

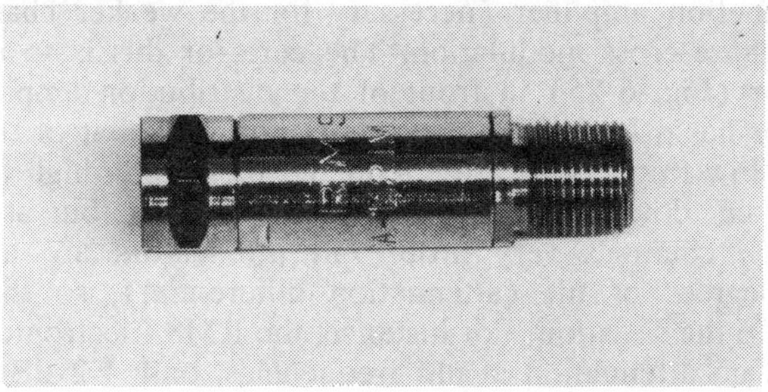

Courtesy Sony Corp. of America

Fig. 6-24. RMS Electronics CA-1121M barrel attenuator.

The RMS Electronics CA-1121M series barrel attenuators (shown in Fig. 6-24) are high-performance 75-ohm coaxial attenuators, available in values of 3, 6, 10, and 20 dB. As Table 6-9 shows, the performance of each of these units is superb and consistent at all tv frequencies. Because of the male F-connector on one end, and the female on the other, this type of attenuator needs no extra cable for insertion into a circuit. Just disconnect the antenna cable from wherever it first connects to, and screw in a 20-dB attenuator.

Attenuators are also used to improve the impedance match (swr) of a filter or tv set to eliminate the possibility of a mismatch-induced ghost. Only 6 dB attenuation is sufficient to reduce a horribly high swr to below 1.7:1.

Tilt Attenuator

The attenuation figures in Chapter 5 shows that high frequencies are attenuated more than low frequencies by transmission lines. This means that even if the signals at the antenna are fairly equal in strength, the lower channels will be strong and the high channels weak after a very long line run. This condition can be aggravated to the problem point if the low channels are much stronger than the high channels to begin with. The result is extremely strong low channels at the output of the transmission line, which will overload

Table 6-9. Measured Performance of 75-Ohm Attenuators

Type of Device	Model	Insertion Loss (dB) 50	200	550	Standing-Wave Ratio 50	200	550	800 (MHz)
Attenuator Pad	RMS CA-1121-3	3.0	3.0	3.2	1.02	1.02	1.04	1.09
	RMS CA-1121-6	6.0	6.0	6.0	1.01	1.01	1.10	1.17
	RMS CA-1121-10	9.3	9.2	9.1	1.01	1.03	1.12	1.17
	RMS CA-1121-20	19.5	19.3	18.5	1.02	1.06	1.17	1.20
Tilt Attenuator	RMS CA-2200	4.3	1.2	0.0	1.05	1.09	1.17	1.32

the distribution amplifier (necessary for the weaker channels) and thereby cause cross modulation. The cure for this is to insert a tilt attenuator (Fig. 6-25) in front of the distribution amplifier. A tilt attenuator has a frequency response opposite to that of transmission line, i.e., low frequencies are attenuated more than high frequencies. This will cut down the strong low channels a lot, but attenuate the weak high channels very little. Tilt attenuators are available in several degrees of tilt (attenuation differential), so they can be matched to the situation. For instance, the RMS Electronics CA-2200 produced no attenuation at uhf frequencies, only 1.2 dB loss in the vhf high band, but 4.3 dB loss at the bottom of the vhf low band. This unit would be selected if the vhf high channels were about 5 dB stronger than the vhf high channels, and the uhf channels were very weak (as they usually are). A higher-value tilt attenuator would be used if the low channels were stronger still.

MISCELLANEOUS

The following items are strictly mechanical devices that provide means for connecting cables together and/or to other devices. However, their construction must be precise in certain ways to avoid impedance irregularities that could cause signal loss or a ghost. Standing-wave ratio is the determining factor here, as acceptable fittings have no measurable insertion loss.

F-Connectors

F-connectors are made in a variety of sizes, to match the cables most often used in matv and catv systems. The F-59 connector fits RG59U-sized cables, the F-56 fits RG6U-sized cables, and the F-11 fits RG11U-sized cables. There is only a small size difference between the F-59 and F-56 connectors, so it is easy to install (or try to install) the wrong connector. The F-59 connector has a bushing

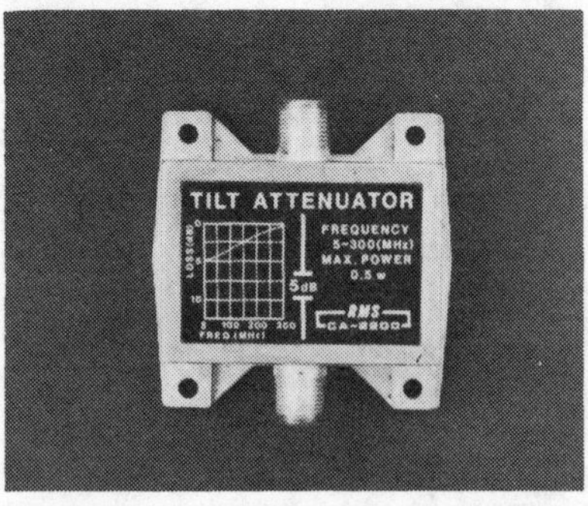

Fig. 6-25. RMS Electronics CA-2200 tilt attenuator.

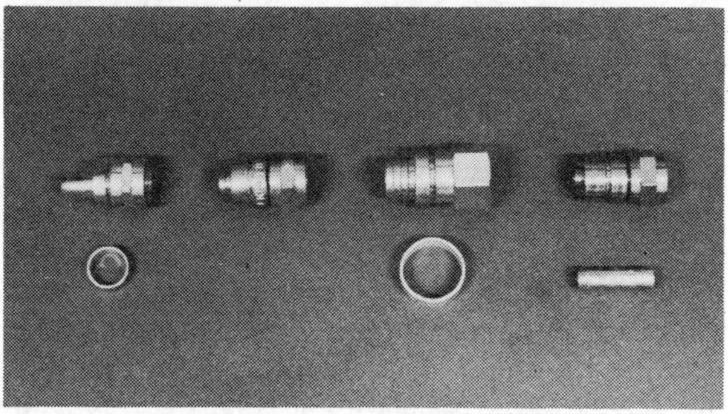

Fig. 6-26. Various types of F-connectors.

hole too small for RG6U-sized cables, whereas an F-56 connector fits too loosely on RG59-sized cables.

F-connectors come in various styles; the ones shown in Fig. 6-26 are typical of the recommended style. The standard F-59 connector shown at the far left has a recess built in that covers the end of the cable after the separate ferrule (crimp ring) is installed. This gives the cable a neat appearance and prevents the braid or drain-wire ends from scratching the fingers when attaching the fitted cable to some device. The Jerrold F-59A shown next has a built-in ferrule. The inner-right connector has the same construction as the standard F-59 (far left), but is intended for giant RG11-sized cable. Notice that this connector has multiple serrations on its bushing; this style is becoming increasingly popular because of superior gripping action on foil-shield cables and easier insertion.

Quality F-connectors also have a constant-diameter bushing hole that ends at a flat surface inside the connector. This results in lowest swr and a watertight seal when the connector is screwed onto the mating female, an important consideration for connectors used outdoors. Companies such as AVA, Finco, Jerrold, RCA, and Winegard supply F-connectors with all of the previous qualities under various model numbers.

Universal connectors are available that can be attached to either RG59U- or RG6U-sized cable. Some have a split bushing (made of sheet metal) that can be adjusted with a special tool to fit either size cable. However, this kind of fitting does not provide a watertight seal, nor will it install as smoothly as normal F-connectors. Because of this latter problem, this type of "universal" F-connector is not recommended for foil-shield cables. RCA supplies an interesting "universal" F-connector with some of their devices. This connector (shown at the far right of Fig. 6-26) is essentially an F-56 connector with built-in ferrule. However, it also includes a size reducer that allows the smaller RG59-sized cable to snugly fit the bushing hole by reducing the hole diameter. Most RG59-sized coax will have a jacket and braid flexible enough to fit over the RG6-sized bushing.

Information on installing F-connectors is given in Chapter 10.

Terminators

Unused outputs on a splitter, and the output port of the last directional coupler in a trunk line, must be terminated for the system to function properly. Although just a carbon resistor in a threaded shell, the performance of the terminator must be consistent at all frequencies, a characteristic much harder to achieve that most people imagine. The Jerrold TR-75F shown in Fig. 6-27 is a low-swr 75-ohm termination suitable for use up to 900 MHz.

Fig. 6-27. 75-ohm terminator.

Cable Switch

In some systems it is desirable to select the signal from one antenna or another, or to switch the signal to one tv or another. Ordinary switches cannot be used at high frequencies because their construction allows signal to leak from the "off" switch terminals to those "on" and/or cause severe mismatch (high swr) that results in signal loss or ghosts. Specially built cable switches, of which the Jerrold Electronics DCS (Fig. 6-28) is a fine example, are required. The measured isolation between nonconnected terminals of the di-

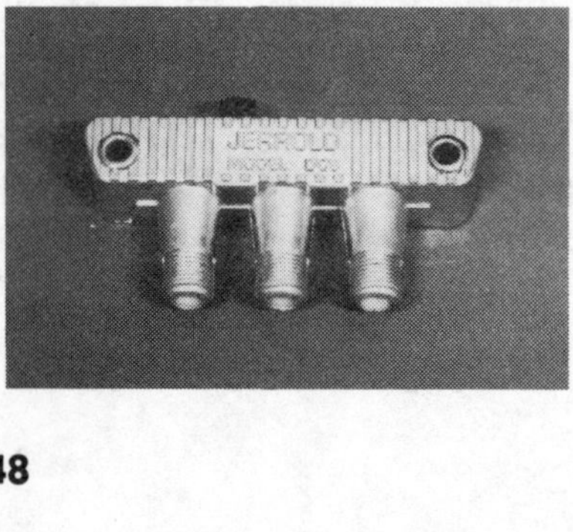

Fig. 6-28. Coaxial cable switch.

rectional couplers run from 80 dB at 550 MHz to over 95 dB at 50 MHz. Moreover, very little signal is lost between connected contacts, the insertion loss ranging from nonmeasurable at 50 MHz to only 0.4 dB at 550 MHz.

Cable Splice

A cable splice (left side of Fig. 6-29) does exactly what its name states: it connects two cables (or any two devices) having male F-connectors. For vhf work with RG59-sized cables, nearly any F-81 type splice will do. For uhf work, low swr is necessary to avoid

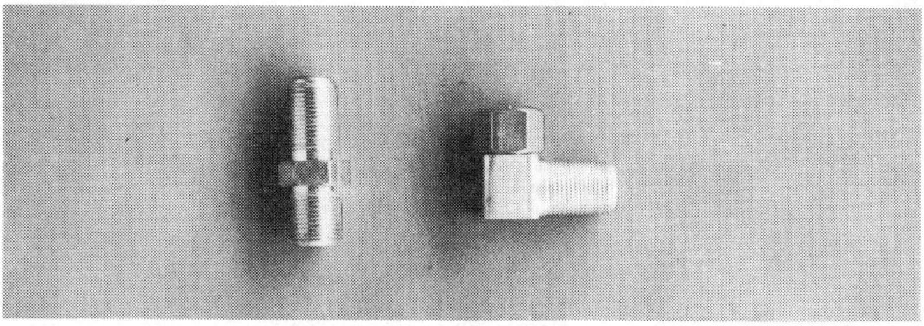

Fig. 6-29. Cable splice and right-angle connector.

signal loss, and for any system using RG6-sized cable the splice must have entry holes big enough to accommodate No. 18 wire. The RMS Electronics 1203 and Color Craft A944 splices have extremely low swr from 50 to 900 MHz, and are nicely made. The RMS splice has the added advantage of having a fully cylindrical body (no flats); hence it is easier to weatherproof for outdoor use.

For the rare consumer installation using RG11-sized cable for the main line, a Winegard F-81M splice is needed to accommodate the No. 14 center conductor of this size cable.

Right-Angle Fitting

The RMS Electronics CA-1208 (right side of Fig. 6-29) is a convenience fitting that allows a cable to be connected at a sharp

Table 6-10. Measured SWR of Fittings and Other Components

Type of Device	Model	Standing-Wave Ratio 50	200	550	800 (MHz)
Cable Splice	RMS F-81	1.01	1.02	1.08	1.13
Right-Angle Fitting	RMS CA-1208	1.01	1.02	1.08	1.12
75-Ω Termination	Jerrold TR-75F	1.01	1.01	1.02	1.04
Cable Switch	Jerrold DCS	1.05	1.15	1.40	1.48

right angle to an F-female. This is very useful when there is insufficient space (or insufficient cable) to allow looping the cable to accommodate the change of direction. The CA-1208 also permits connecting RG11-sized cable to components whose F-females have entry holes too small to accept a No. 14 center conductor.

The same criteria apply to this device as in the case of the cable splice. As the swr figures in Table 6-10 show, the CA-1208 has extremely low swr, in addition to nonmeasurable insertion loss.

CHAPTER 7

Antenna Hardware

To produce an effective antenna system, a good antenna must be supported by properly selected and installed hardware. This includes a mast or tower to support the antenna, mounting brackets to secure the mast to the habitation, guying materials to brace the mast or tower, rotors to vary the antenna orientation, and lightning protection devices. The antenna hardware itself is described in this chapter. How to install the various items in your antenna system is described in general terms in the final chapter. Exact installation information on some devices must be obtained from the instruction sheets packed with them.

MASTS AND TOWERS

The most common means of supporting a tv antenna is a 10-foot length of 1¼-inch-diameter steel tubing called a *mast*. This mast may be a single 10-foot section, or made from a pair of 5-foot sections. Most mast sections these days are made like the one shown at the right in Fig. 7-1. A 6-inch length on one end is reduced in size (swaged) so that end will fit tightly inside the plain end of another mast section. This allows tall masts to be built from relatively short (hence easily transportable) sections. These mast sections are available in 16- 18-, and 20-gauge wall thicknesses. A few manufacturers also offer 1½-inch-diameter masts, but these are rarely stocked at consumer outlets.

With wide-spaced and securely anchored wall mounts, a single 10-foot mast section will safely support vhf or vhf-uhf antennas having boom lengths of up to 120 inches in moderately strong winds. A

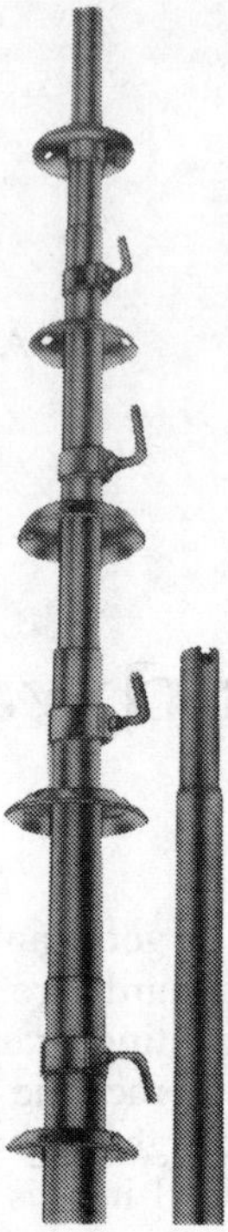

DURATUBE TELESCOPING MAST

SPECIAL HEAVY-GAUGE STEEL.
50% STRONGER THAN HEAVY-GAUGE,
HIGH-CARBON STEEL.

MODEL	DESCRIPTION
1020	20 FT
1030	30 FT
1040	40 FT
1050	50 FT

DURATUBE STRAIGHT LENGTH		
1607	5 FT	1-1/4"x16 GA
1612	10 FT	1-1/4"x16 GA
1807	5 FT	1-1/4"x18 GA
1812	10 FT	1-1/4"x18 GA
2007	5 FT	1-1/4"x20 GA
2012	10 FT	1-1/4"x20 GA

Courtesy Channel Master

Fig. 7-1. Standard and telescoping mast sections.

mast consisting of a 16-gauge 10-foot lower section and 20-gauge 5-foot upper section will safely support an antenna with a 70-inch boom under the same conditions. Taller masts, or those supporting bigger antennas, *must be* guyed (braced with taut wires) as described in Chapter 10.

Guyed masts can be built from 10-foot mast sections and the guying hardware described in this chapter. However, the most convenient way is to buy a telescoping mast like that shown at the left in Fig. 7-1. These masts consist of mast sections of progressively smaller diameter that nest for transportation. At the antenna site the sections are pulled out and locked in place to form the extended mast. Integral guy rings eliminate bothering with this detail.

Telescoping masts are available in lengths ranging from 20 to 50 feet. Some companies also offer a choice of lightweight or heavy-duty masts in each length. The listing at the bottom of Fig. 7-1 shows the wall thickness (gauge) of each section in Channel Master's lightweight (2000), medium-weight (1800) and heavy-duty (1600) series of telescoping masts. Their Duratube® masts and mast sections

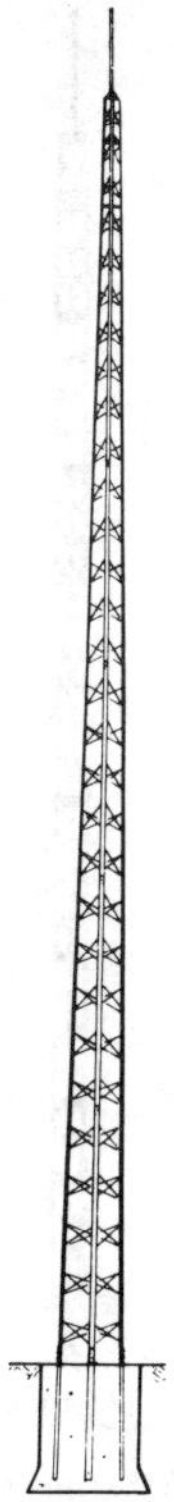

(*A*) *QDMX-52 tower.*

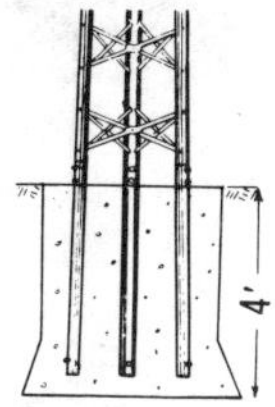

(*B*) *Concrete base.*

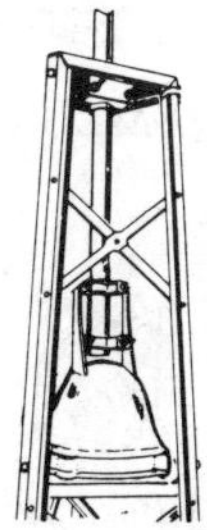

(*C*) *Rotor mounting.*

QBBMB

(*D*) *Bearing in tower top plate.*

Courtesy TACO/Jerrold

Fig. 7-2. Jerrold QDMX self-supporting tower.

are featured here because of their exceptionally good rustproofing and high strength.

When considerable antenna height is required, and guying is not possible, the answer is the self-supporting or house-supported tower. The self-supporting tower is a rigid and (usually) tapered latticework structure (Fig. 7-2A) that is set into a deep concrete base in the earth (Fig. 7-2B). The antenna is attached to a mast section atop the tower. Most towers have provisions for mounting an antenna rotator near the top of the tower (Fig. 7-2C). A bearing (Fig. 7-2D) in the tower top plate through which the mast passes eliminates stress on the rotator. The Jerrold QDMX tower is a good example of the self-supporting tower. It is shipped in 8-foot sections that conveniently nest for shipment. Heights of 28 to 68 feet above ground are available.

The house-supported or "assisted" tower (Fig. 7-3A) uses a base plate staked to the earth (Fig. 7-3B) and a bracket fastened to the

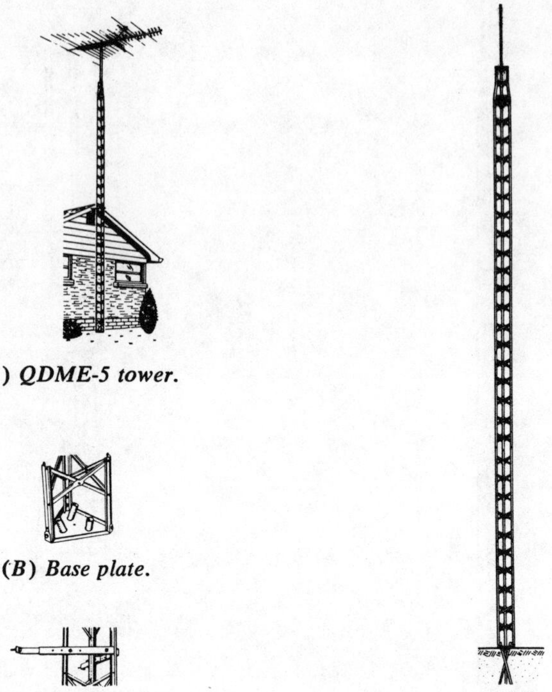

(*A*) *QDME-5 tower.*

(*B*) *Base plate.*

(*C*) *Bracket.*

(*D*) *Structure.*

Courtesy TACO/Jerrold

Fig. 7-3. Jerrold QDME house-supported tower.

upper part of the house (Fig. 7-3C) for support. Except for these points and the straight sides (constant cross section), these towers are much the same as the self-supporting towers. The Jerrold QDME (Fig. 7-3D) is a good example of this type. Tower heights of up to 52 feet are practical.

MOUNTING BRACKETS AND DEVICES

There are many ways to attach antenna masts atop a building. These include tripod mounts, base mounts, wall mounts, and chimney mounts. Each has its advantages and applications, as described in the following paragraphs.

Tripod Mounts

Tripod mounts are halfway between true towers and simple mounting brackets. The universal tripod mount (Fig. 7-4) is actually a 5-foot mast section with built-in mounting fixtures. The pivoting leg

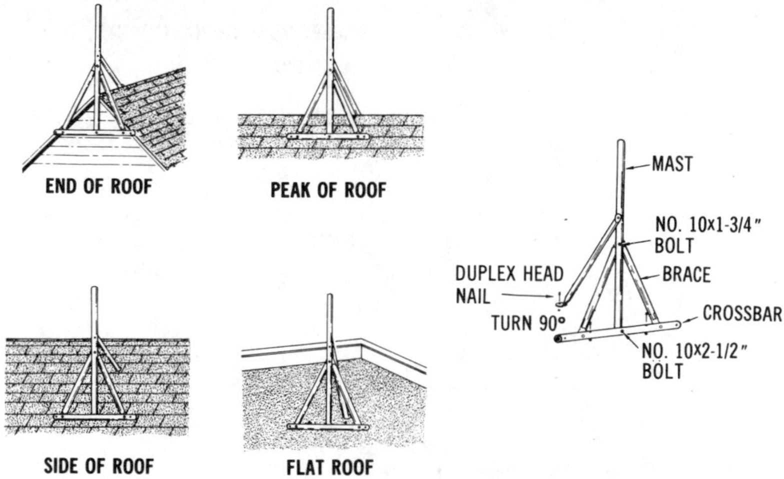

Courtesy Gavin Electronics

Fig. 7-4. Universal tripod mount and mounting options.

can be arranged in any of several ways to allow the mounting techniques shown. A 5-foot mast section with swaged end is inserted in the vertical member of the mount to hold the antenna.

The classic tripod mount is shown in Fig. 7-5. This is strictly a mount; a mast must be inserted. This tripod mount is intended primarily for flat roofs, although the Channel Master Models 9137 and 9138 have adjustable feet that allow the tripod to straddle a peaked roof. A 3-foot-high tripod will support the usual 10-foot mast and medium-sized antenna easily. If the mounting bolts are securely set into the roof, the 5-foot-high heavy-duty models will support an unguyed 15-foot mast as described earlier.

Base Mounts

The base mounts shown in Fig. 7-6 merely hold the bottom end of the mast in place; guy wires must be used to keep the mast from tipping over. Most of these mounts are adaptable to flat or peaked roofs. A few, like the 9005 and 9001, are intended solely for a specific application.

Chimney Mounts

Chimney mounts (Fig. 7-7) are the easiest to install of all the mounting devices. This is because no drilling is needed; a pair of mounts are held to the corner of the chimney by straps that encircle the chimney. The spacing between mounts should be somewhere in the 20- to 30-inch region, the wider the better. However, the dimensions of the chimney often dictate the maximum possible spacing.

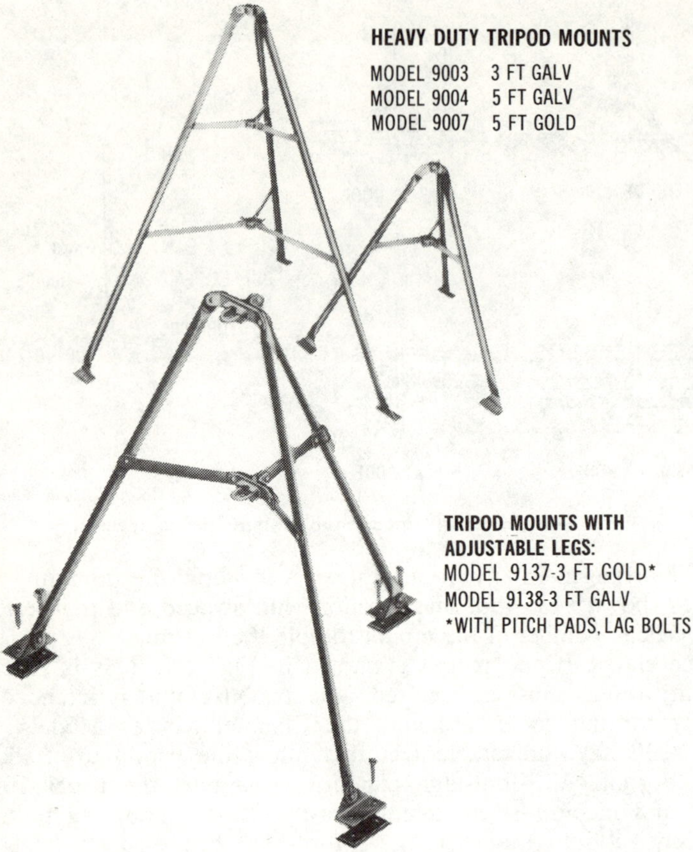

Courtesy Channel Master

Fig. 7-5. Channel Master tripod mounts.

Chimney mounting has a serious drawback; the effluent from an active chimney is very corrosive. An antenna that will last 10 years in a smoke-free site might last only 3 years atop a chimney. If a mast-mounted preamplifier is used on a chimney-mounted antenna, the result could be a financial disaster. The effects can be mitigated by mounting the antenna as high as possible above the chimney, so don't use less than a 10-foot mast section. In fact, if the chimney is large and in good condition, and heavy-duty mounts are used, a 15-foot mast as described earlier might be a good idea.

Wall Mounts

Wall mounts (Fig. 7-8) are also used in pairs to support a mast. If properly anchored, they offer about the sturdiest mount available for an unguyed mast. The spacing between mounts should be in the 20- to 30-inch range, the wider the better. When mounted on the outside wall of a wooden house, the upper of the two mounts is set close to the roof. If there is considerable roof overhang, long (and well-braced) mounts like the Channel Master 9034–9036 series must be used so the mast will clear the roof edge. If the overhang is considerable, an eave mount is advisable if the roof-edge construction is sturdy enough. The usual mounting place for wall mounts

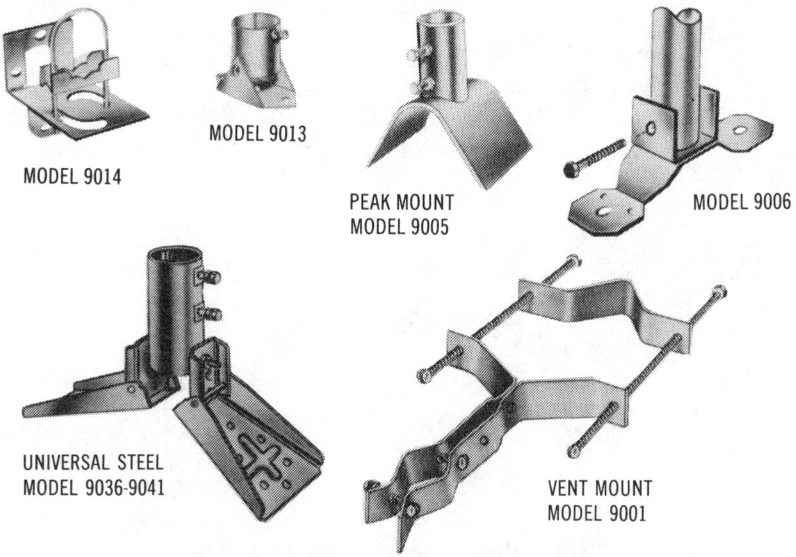

Courtesy Channel Master

Fig. 7-6. Channel Master base mounts.

on brick buildings is the inner side of a parapet wall. This is the mounting technique most commonly used on apartment houses.

Mounting Kits

Many of these mounts are available as part of mounting kits. A mounting kit contains everything necessary to install an antenna: mount, masting, transmission line, and standoffs. A typical chimney-mount kit is shown in Fig. 7-9. These kits are fine for mounting uhf or small vhf antennas in strong-signal areas, but separate components, chosen in size and quantity appropriate to the particular installation, are needed for mounting large antennas in critical installations.

Various mounts are shown installed on a peak-roof house in Fig. 7-10. This illustrates how the different mounts are installed, and how several masts can be erected on one house.

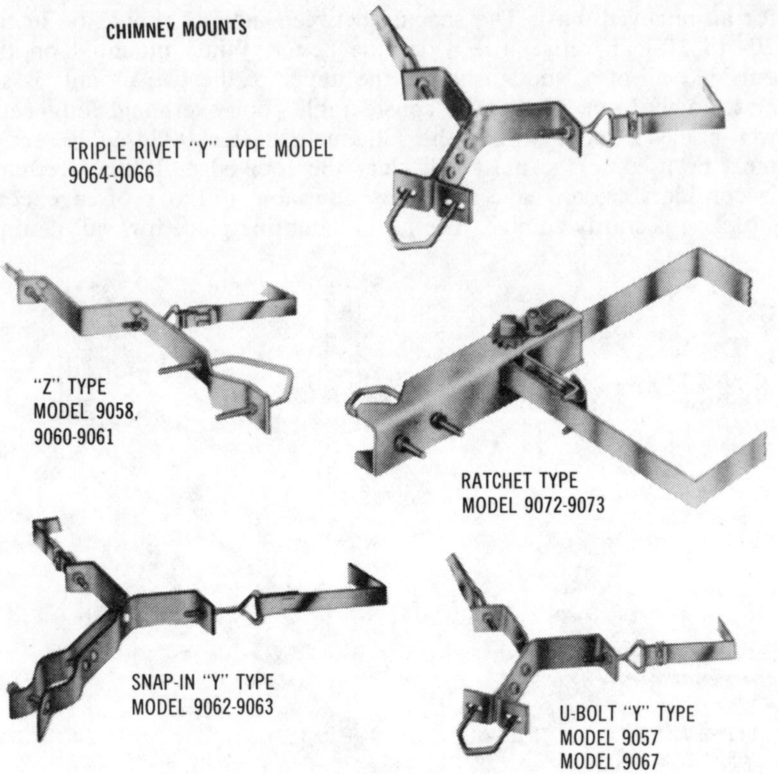

Courtesy Channel Master

Fig. 7-7. Channel Master chimney mounts.

GUYING HARDWARE

Base mounts simply anchor the bottom end of a mast section to the roof; the mast must be guyed to keep it and the antenna from falling over. The guy wire is usually stranded steel wire. The wire size depends on the mast height and antenna size; four strands of No. 20 to six strands of No. 18 are the sizes most commonly used for home and small matv installations.

The antenna ends of the guy wires are fastened to a guy-wire clamp, but a floating guy ring such as the Radio Shack 15-835 (Fig. 7-11) is best. This technique permits the mast to be turned from the ground after erection, to adjust antenna orientation. These rings also allow either a three- or four-wire guy system to be properly attached. The roof ends of the guy wires are attached to guy hooks set into the roof or sidewall. A turnbuckle in each guy wire permits exact positioning of the mast (by adjusting the individual wire

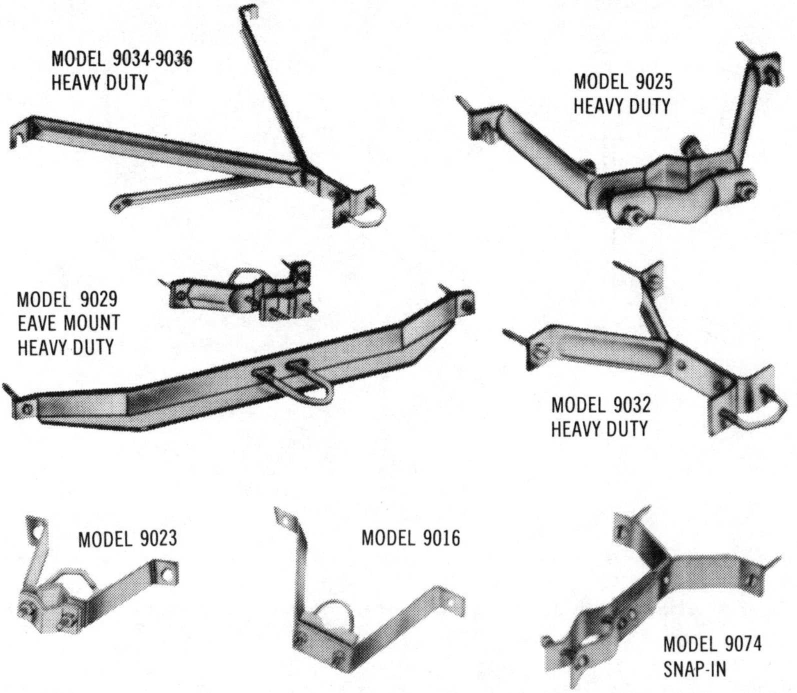

Courtesy Channel Master

Fig. 7-8. Channel Master wall mounts.

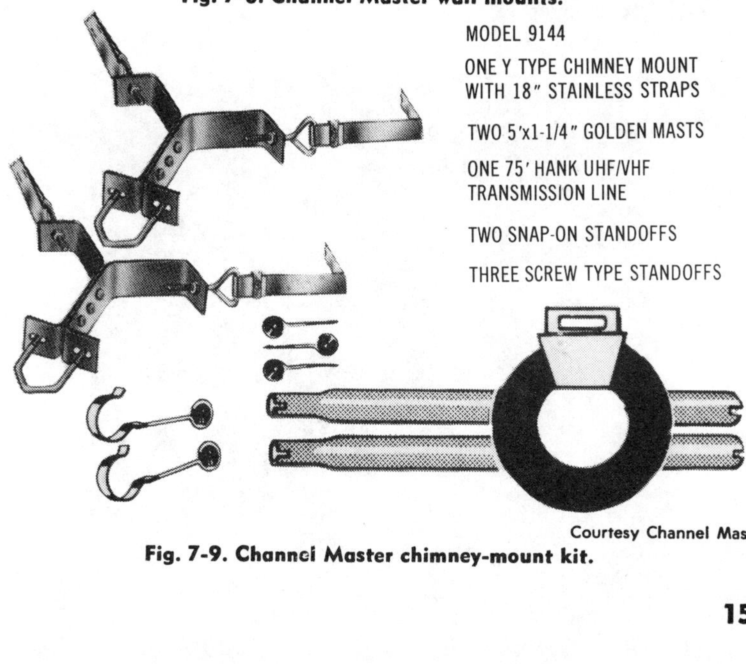

Courtesy Channel Master

Fig. 7-9. Channel Master chimney-mount kit.

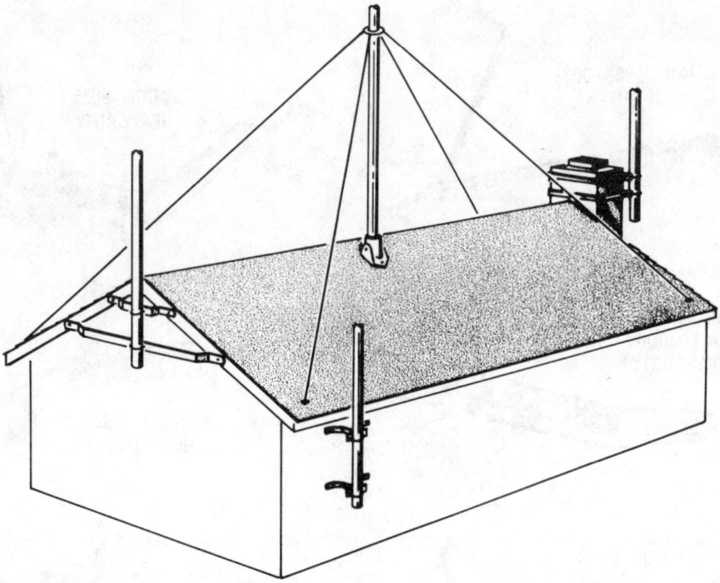

Fig. 7-10. House with eave, roof, chimney, and wall-mounted antenna masts.

lengths) and tension adjustment. The cable clamps and almond-shaped wire thimbles at the far right of Fig. 7-11 are used to fasten the heavier wire sizes.

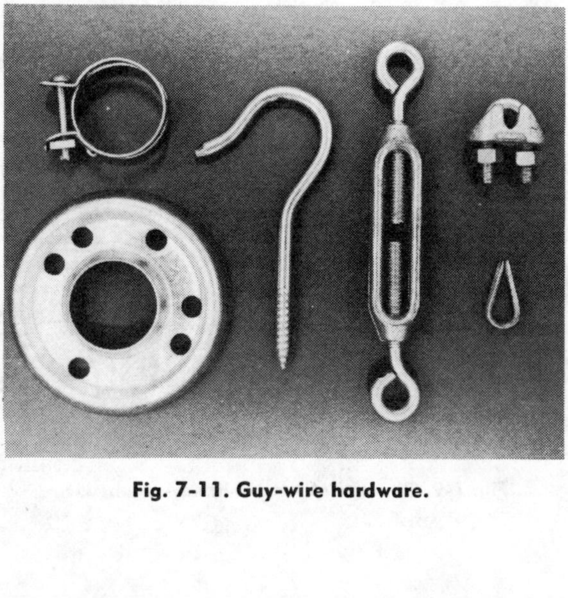

Fig. 7-11. Guy-wire hardware.

STANDOFFS

When unshielded twinlead is used for the transmission line, standoff insulators are needed to hold the line away from the mast, roof, and walls. Moreover, though the insulating effects are not needed, standoffs are also used on occasion to physically anchor shielded lines. A variety of types and sizes of standoffs is available for this purpose; a few representative types are shown in Fig. 7-12. The

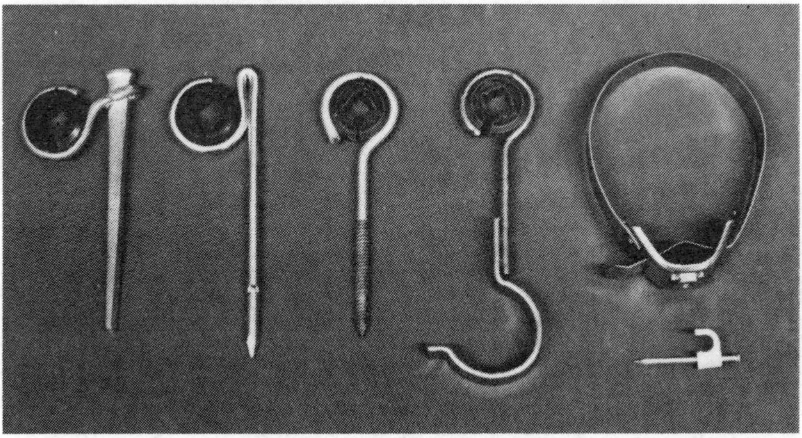

Fig. 7-12. Standoff insulator styles.

popular masonry-nail drive-in standoff at the far left can be driven directly either into cement or into an expansion anchor set in brick. For wooden walls and shingle-covered roofs, either the nail drive-in or screw-thread standoffs can be used. The screw-thread standoff is available in lengths of 3½ to 7½ inches; use the longest version whenever possible. The snap-on standoff at the right is used on standard (1¼-inch-diameter) antenna masts. For larger-diameter masts or other structures use a screw-thread standoff set in a strap mount (far right).

All of the standoffs shown in Fig. 7-12 have what is called the uhf-style polyethylene insert or grommet. This is only one of three basic types shown in Fig. 7-13. Insert (a) is the vhf style; it is intended for either standard flat twinlead or RG59-sized coaxial cable. Insert (b) is a uhf style; it will hold either standard flat twinlead or tubular twinlead (such as Belden 8275). Insert (c) is one of the few available that can hold large-size foam-encased twinlead (Belden 8285) or shielded twinlead (Belden 8290). Doing this requires a large oval cutout, such as found in IE Manufacturing Company's "Universal" insert.

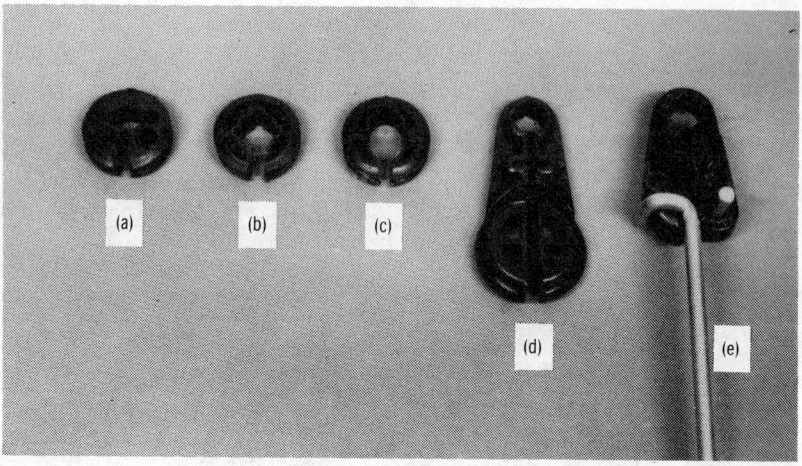

Fig. 7-13. Standoff insert styles.

All of the previous inserts position the line within the wire ring holding the insert. The types shown in Fig. 7-13 (d) and (e) hold the line outside the metal portion of the standoff. This reduces the amount of impedance irregularity caused by the metal ring, and holds the line further away from the surface or object the standoff is mounted on. Insert (d) fits into standard standoffs like those shown in Fig. 7-12; the insert is held by a wire ring. This style insert will hold tubular twinlead or lightweight flat twinlead. Standoffs of various styles are available from a few manufacturers with this insert. The inserts can also be purchased separately (Lafayette 18-0177) and substituted for the regular inserts in standoffs already installed, to improve the system. Insert (e) is similar in operation to insert (d), but mounts differently. Line-Lok® standoffs using insert (e) are available in screw thread, snap-on, and strap-mount styles from IE Manufacturing Co.

Coaxial cable and rotator cable (when used) require no insulators, but must be secured somehow to prevent abrasion from roof edges and for neat inside installations. They should be taped to the antenna mast and pipes, but fastening devices are needed when the cable traverses a wide surface, such as a wall. For brick or cement-block walls, use a masonry-nail standoff with vhf grommet for rotator cable and RG59-sized coax; RG6-sized coax is best accommodated by the uhf grommet. For wood surfaces, particularly indoors, a plastic cable clamp like the one shown at bottom right in Fig. 7-12 permits neat and easy installation of coax. These clamps are available in light or dark colors and three sizes. RMS Electronics Models 1170/1169 fit RG59-sized cable, Models 1172/1171 fit RG6-sized cable, and Models 1174/1173 fit RG11.

LIGHTNING-PROTECTION DEVICES

An antenna mast is a tempting target for lightning, particularly when the mast is the highest structure in the vicinity. If precautions are not taken, the earth-sky lightning path will be completed through the transmission line, tv power cord, and house wiring. At best the tv will be destroyed, at worst the entire house will burn. The danger and damage can be minimized by techniques (explained in detail in Chapter 10) using lightning-protection devices such as ground wire and ground rods. No. 8 aluminum wire is the most commonly used ground wire; it is sold nearly everywhere in 50- to 1000-foot spools. The ground rod should be at least 4 feet long, and heavily copper plated so it won't rust away.

Coaxial cable must also be grounded. The Winegard F-81GB grounding block at the left of Fig. 7-14 is made for this purpose; it is essentially an F-81 cable splice set in a block to which a ground wire can be attached.

Direct-stroke protection of unshielded twinlead is not possible, but a lightning "arrester" like the JFD Electronics AT131S shown in Fig. 7-14 will drain static charges off the line.

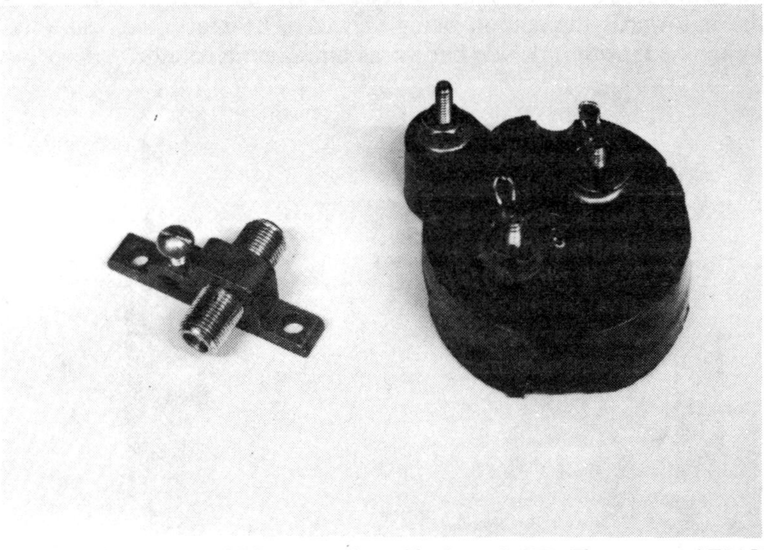

Fig. 7-14. Winegard F-81GB grounding block and JFD Electronics AT131S lightning arrester.

ANTENNA ROTATORS

If only one station is receivable, or all stations transmit from the same site (as in New York and Chicago), an antenna rotator is not

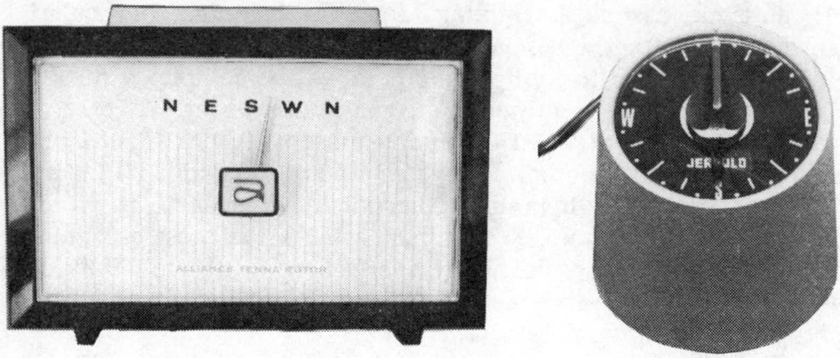

(*A*) *Manual.* (*B*) *Automatic.*

Fig. 7-15. Manual and automatic rotator control boxes.

needed. However, if a number of stations are receivable, they lie in different directions from your reception site, and one broadband antenna is used, a device is necessary to turn the antenna and precisely aim it towards the station being viewed. The most practical way of doing this is with a device known as an antenna rotator.

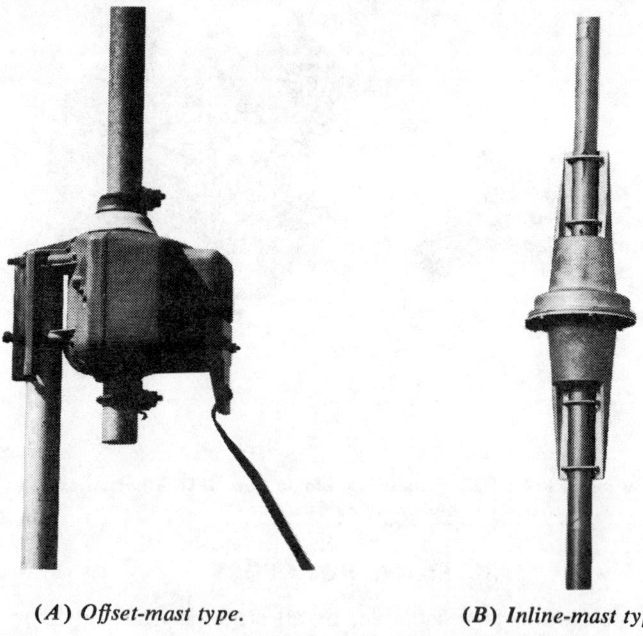

(*A*) *Offset-mast type.* (*B*) *Inline-mast type.*

Fig. 7-16. Antenna-rotator motor units.

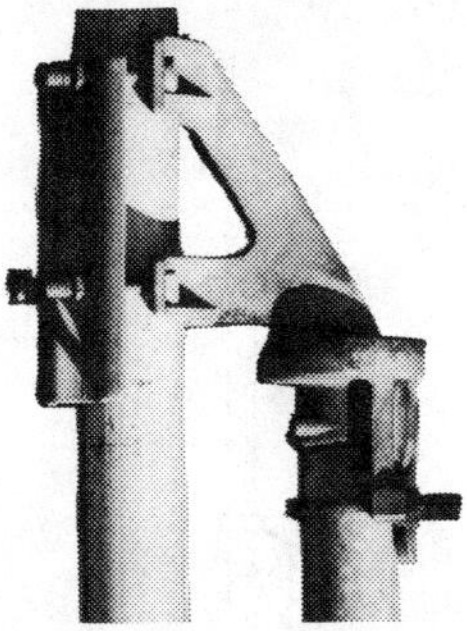

(*A*) *Thrust bearing.*

(*B*) *Alignment bearing.*

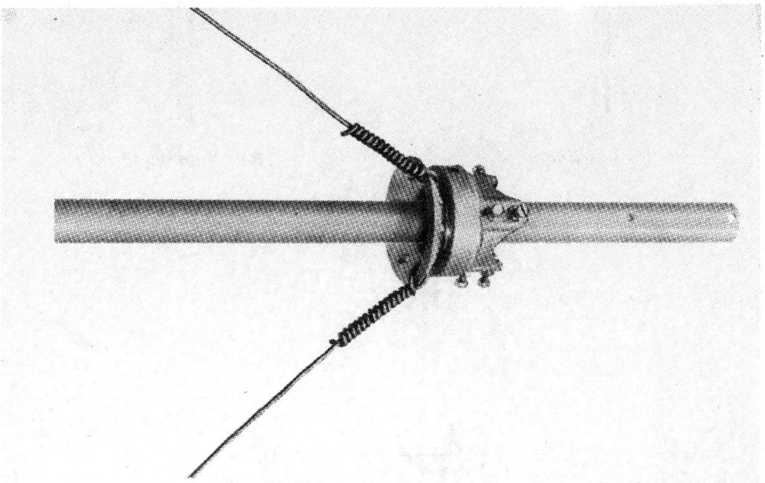

(*C*) *Rotating guy ring.*

Courtesy Channel Master

Fig. 7-17. Rotated-mast supports.

The antenna rotator consists of a remotely controlled electric motor and transmission that mounts atop the antenna mast, a control box with direction indicator that is located near the tv set, and sufficient control cable to connect the two. The cable may contain anywhere from three to five conductors, the exact number depending on the motor-drive indicator systems.

Rotator operation depends on the type of control box used. If the control box is the manual type, like the Alliance T-45 shown in Fig. 7-15A, the antenna is rotated in one direction for as long as

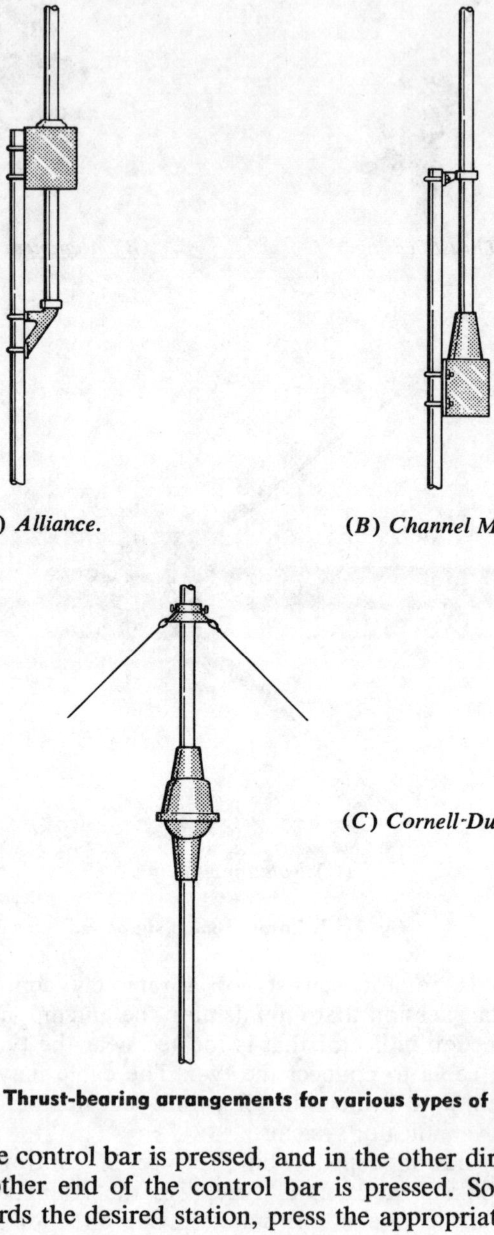

(*A*) *Alliance.* (*B*) *Channel Master.*

(*C*) *Cornell-Dubilier.*

Fig. 7-18. Thrust-bearing arrangements for various types of rotators.

one end of the control bar is pressed, and in the other direction for as long as the other end of the control bar is pressed. So, to aim the antenna towards the desired station, press the appropriate end of the control bar atop the unit, and hold it down until the pointer indicates the direction of the station.

If the control box is the automatic type, like the TACO unit shown in Fig. 7-15B, simply turn the direction control to the bearing of the desired station. The antenna will rotate to the proper direction and automatically stop at that point. Automatic units are, naturally, the most expensive type.

The motor units of a particular manufacturer are often the same regardless of the type of control unit. Motor units are geared so most require one minute to rotate 360°. Most are sturdy enough to handle a 10-foot antenna in fairly high winds. For larger antennas, especially in very windy and icy climates, a heavy-duty motor unit is recommended. Channel Master and Cornell-Dubilier have such units available.

The main difference in motor units is the positioning of the fixed (lower) mast and the rotated (upper) mast. The masts are offset in the very popular Alliance rotators (Fig. 7-16A), the Channel Master, and the small Cornell-Dubilier rotators. On the other hand, the big Cornells and the TACO (Fig. 7-16B) position the masts in-line. Each type has its advantages and disadvantages. The offset type permits easy use of a thrust bearing (Fig. 7-17A). A thrust bearing is needed when the antenna is very heavy and/or the rotated mast section must be long (as when two or more antennas are mounted on that mast section). When the rotated-mast section passes through the motor unit (as in Fig. 7-18A), the thrust bearing is located *below* the motor unit. When the rotated mast does *not* pass through, an alignment bearing (Fig. 7-17B) is mounted *above* the motor unit (Fig. 7-18B).

An in-line rotator will support a heavy antenna without the need of a thrust bearing *provided* the upper mast is very short (one to two feet). This imposes a *one-antenna* limitation on the mast. In-line types require a rotating guy ring (Fig. 7-17C) located about two feet above the rotator (Fig. 7-18C) to permit using a long upper mast.

CHAPTER 8

Signal Optimization Techniques

When the signal level at the reception site is very low, or ghosts or interference mar reception, certain techniques can be employed to optimize the signal or at least improve it. This chapter details methods of increasing signal level, preventing the degradation of signal-to-noise ratio, and eliminating ghosts and interfering signals. Various methods of receiving stations that lie in different directions are also discussed.

INCREASED SIGNAL OUTPUT

If an existing antenna provides inadequate output, or the high-gain types recommended in Chapters 2–4 appear unlikely to do the job because of extreme reception distance or problem terrain, one or more gain-improvement techniques are required.

Larger Antenna

A larger antenna is theoretically the best way to increase the signal provided to the tv receiver or signal-distribution system, because the increased signal is obtained without degrading the signal-to-noise ratio. However, there are certain "hitches" to this technique.

The average vhf or vhf-uhf antenna has a boom length of 70–80 inches. Since antennas with boom lengths of 150–200 inches are available, moving up to a bigger antenna is usually practical. However, the point to remember when a particular model proves insufficient is that getting the next model up in the line will not produce a noticeable gain improvement. To show useful improvement, the bigger antenna must be at least twice the size of the one it is replacing. The reason why becomes obvious by comparing the gain and physical specifications for the Winegard CH-4052 and CH-5200, as is done in Table 8-1. The average gain increase obtained at nearly triple the boom length and over twice the weight amounts to 2 dB on the low band and 3.2 dB on the high band. This is typical of many other vhf and vhf-uhf antenna lines. The gain difference between any two consecutive models in a full-sized antenna line is only a fraction of a decibel.

Table 8-1. Comparison of Medium-Sized and Very Large VHF Antennas

Model	Boom Length (in)	Weight (lbs)	List Price ($)	Average Gain (dB) VHF-L	Average Gain (dB) VHF-H	Average Beamwidth (°) VHF-L	Average Beamwidth (°) VHF-H
CH-4052	75	6¼	47	5.2	8.7	71	48
CH-5200	200	14¼	122	7.2	11.9	62	36

The problem with uhf antennas is a little different. Because fairly elaborate uhf antennas can be built with no dimensions more than 60 inches long, most uhf antennas are already fairly close to the maximum practical gain for their design. For example, adding ten more directors to a corner-reflector yagi will produce an average gain increase of only 1½ dB or so, with most of the increase at the high end of the band. Doubling the size of a corner reflector does likewise, but most of the gain increase now is in the lower half of the band. The one uhf antenna type that gives a really worthwhile gain improvement as its size is increased is the parabolic reflector antenna, mainly because the performance of a small parabolic is so poor. Keep in mind, though, that 6–7-foot parabolics (and to a lesser extent long-boom yagis) have several characteristics that make their utilization difficult. They are as heavy as a good-sized vhf antenna, have extremely high wind resistance, and their dimensions are such that these antennas will interfere with the operation of nearby vhf antennas. These considerations require heavy-duty, well-guyed, and well-anchored masts that are located away from vhf antennas.

Accessory Director Arrays

In certain cases the gain of an existing uhf antenna can be increased by what amounts to making it bigger. Several companies make an attachment that adds more directors to the front of one or more of their uhf or vhf-uhf antennas. Jerrold's 8PZ director array (eight elements) fits their VU- and ZIP-series of all-channel antennas. Winegard's CH-0820 has 13 elements on a short (18-inch) boom; it is intended only for six models in its Chromstar line of vhf-uhf antennas. The Antenna Corp. of America has the largest director array, nine wide-spaced directors on a 36-inch boom. This array, the PP-800, fits several all-channel antennas and the AC-316 corner-reflector yagi, on which it is shown in Fig. 8-1.

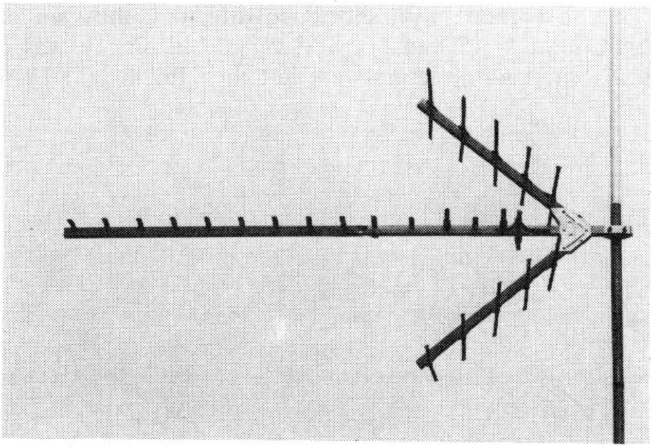

Fig. 8-1. PP-800 director array on Antenna Corp. of America AC-316.

Director arrays have several characteristics of which you should be aware before running out and buying one for your antenna. First, they do not increase the gain evenly across the band. Maximum gain improvement occurs *near* the high end of the band. If the very weak stations are at the low end of the band, the improvement is rarely worth the cost and effort. Second, if the directors are of the same length as those on the antenna proper, the high-end gain will decrease at a more rapid rate (See Fig. 4-10). A third factor is the increased weight and boom length; sometimes it is difficult to keep the mast from bending and thereby cause the antenna to point downward. If the weak stations are located at frequencies where the director array is most beneficial, however, these devices can be a cheap and efficient way to boost the signal.

Table 8-2 shows the gain improvement produced by adding a PP-800 to a modified AC-316. (The author had modified the corner reflector for an unrelated experiment, so the data presented here is simply an indication of the effects of the PP-800.) Note how the gain difference is relatively small (around 1 dB) at the low end of the uhf band but gradually increases to substantial amounts (around 3 dB) over Channels 50–75. Note also that the greatest improvement occurs just before the gain nosedives at Channel 80. Comparing the polar patterns (Figs. 8-2A and 8-2B) shows that the −3-dB beamwidth narrows only 5 percent at the low end of the band (from 56° to 53°), but decreases by 18 percent at the high end (from 36° to 30½°).

These results were obtained with the directors on both the antenna and director array shortened to 5½ inches for maximum frequency coverage. When a PP-800 with full-length (7-inch) directors was installed on an antenna with shortened directors, the gain peak occurred on Channels 45 and 50, and very little pickup was possible by Channel 65. If all directors were left their full length (7 inches),

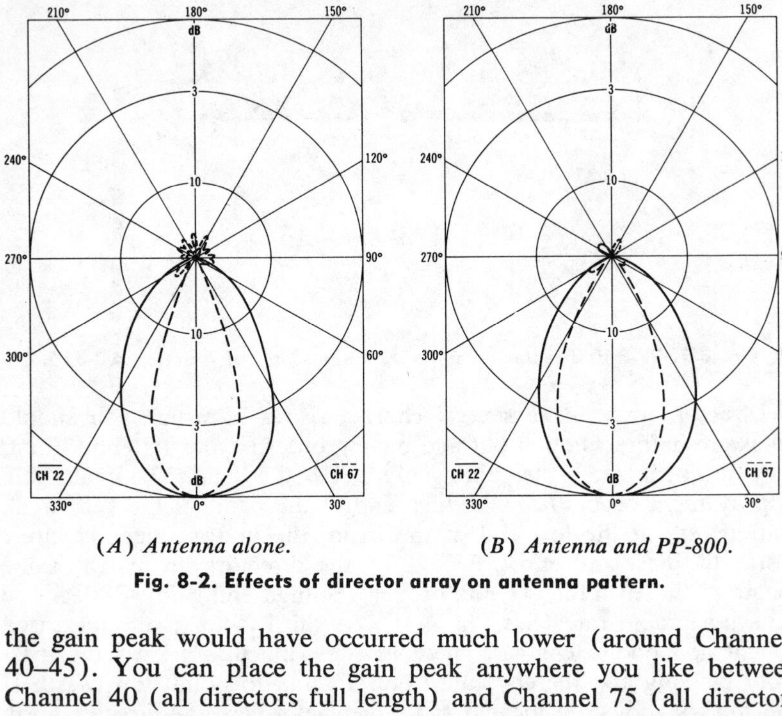

(*A*) *Antenna alone.* (*B*) *Antenna and PP-800.*

Fig. 8-2. Effects of director array on antenna pattern.

the gain peak would have occurred much lower (around Channels 40–45). You can place the gain peak anywhere you like between Channel 40 (all directors full length) and Channel 75 (all directors shortened) by the simple process of shortening just the right number of directors. Starting with all directors full length, shorten them one at a time, while monitoring the antenna output on the desired channel with a field-strength meter (see Chapter 9). Start with the frontmost director. Stop shortening directors as soon as no further improvement

Table 8-2. Gain Improvement vs. Frequency of A.C.A. PP-800

Channel	22	25	30	35	40	45	50	55	60	65	70	75	80
dB	0.8	0.9	1.1	1.2	1.2	1.3	2.3	2.6	2.5	2.7	3.1	3.1	−0.8

results, or if the gain starts to decrease. Be certain to do this at a time when the signal is steady.

Higher Antennas

Significantly greater signal output can *usually* be attained by simply increasing the antenna height. This technique is most dependable in locations where the increased height will put the antenna "in the clear." For instance, if all the surrounding buildings and trees are the same height as or slightly higher than your own location, going from a 10-foot roof-mounted mast to a 30-foot mast will usually produce a very large increase in signal strength, larger by far than is likely to be achieved by using a bigger antenna on the 10-foot mast.

Literature prepared by Jerrold representatives claims that *doubling the antenna height* (above ground) increases the output voltage of a vhf antenna by 6 dB, the output of a uhf antenna by 12 dB.* In some circumstances, however, increased antenna height will *reduce* the signal level. This results from the fact that the output of an antenna is always the sum of the direct signal and reflected signals. The classic interference pattern of direct and reflected signals produces a signal profile that alternately increases and decreases with altitude at distances relatively close to the ground or a large reflecting surface (such as a metal roof). This means that although the average signal strength will increase with increasing height, there are spots where it will decrease as the antenna is raised. This effect occurs at all tv frequencies, but because the distance between signal "hot spots" (nodes) and "cold spots" (nulls) is related to wavelength, dramatic changes can occur within a foot or two at uhf frequencies. This is why it is important to erect temporary masts at critical installations and look for the antenna location and height that gives the best results on the difficult channels. Once a hot spot has been located, a permanent installation can be erected.

Peculiar variations in signal strength with location and height may occur when the antenna is surrounded by buildings too tall for a higher mast to clear. Fortunately, cases like this are most common in metropolitan areas, where the signal strength is no problem, and least ghosting in the picture is the criterion for antenna height and location.

The effect of increased antenna height at a typical rooftop location is shown in Table 8-3. These measurements were made on a roof 25 feet above street level, so the total antenna height is given in parenthesis. Notice the often large increases in low-band vhf and uhf signal levels for a modest percentage change in antenna height, and the virtually "no change" situation in the vhf high-band signal levels. The large improvement in low-band vhf output occurred because the

---

* Bert Wolf, "Towers—R_x for Fringe Areas," *Electronic Servicing,* May 1977.

8- and 9½-foot spacing between antenna and roof is so little (in terms of wavelength) on the low band that the antenna's pickup patterns were effected by the proximity of the roof. The additional spacing placed the antenna far enough away to make the most of the signal levels available. However, on the high band 8- to 9½-foot separation is enough for the antennas to work properly, so the additional 8 feet of height has little effect. Instead, the signal strength increases *and* decreases slightly as the antenna height changes. Further measurement of Channel 13 output versus height showed a 2-dB variation in output as the antenna was slowly raised from 9½ to 18 feet above roof level. The greatest output in this instance occurred at 12½ feet from the roof. The same experiment on Channel 4 showed a steady increase in output up to 17 feet.

Separate Antennas

In Chapter 2 it was shown how a single-channel antenna produced much more gain on that channel than could a broadband (12-channel) antenna of the same size. Similarly, in Chapter 3 it was shown how a limited-range (16-channel) corner-reflector yagi produced about 4½ dB higher average gain than the broadband (56-channel) version. Separate or specialized antennas offer a practical and economical method of achieving maximum antenna gain when just a few channels are receivable, and circumstances are right.

The use of a separate antenna for each channel is widely employed at catv and large matv systems. Not only does this provide high gain and clean patterns for each channel, but also makes all channels simultaneously available because no rotation is needed. The output of each antenna can also be amplified or attenuated as necessary before combining to produce equal signal level on each channel. Systems of this type are described in "MATV HEAD ENDS" at the end of this chapter. The disadvantage of these systems is that they can become expensive and unwieldy if very many channels are involved. Also, the combining devices introduce 1.5–2.5 dB signal attenuation, so much of the gain advantage of using single-channel antennas is lost. However, if only two or three antennas are involved and they are for channels in different bands (uhf, vhf-L, vhf-H), their output can be combined in a nearly lossless manner by using 75-ohm band separators backwards as band combiners. The *antenna selection* criteria for building a practical, high-performance home or small matv-antenna system are as follows:

1. One receivable channel per vhf band—Use a single-channel yagi for each vhf band on which this occurs.
2. Two or more receivable channels on the same vhf band—Use a single-band antenna for each vhf band on which this occurs.

3. One receivable uhf channel—Use a narrow-band antenna covering that channel.
4. Two or three receivable uhf channels close in frequency—Use a narrow-band antenna covering the receivable channels.
5. Two or more receivable uhf channels spread over the band—Use a broadband antenna.

The rules for combining the selected separate antennas are as follows:

1. To combine a vhf high-band and a vhf low-band antenna, use a 75-ohm vhf H/L band separator backwards as a combiner, as in Fig. 8-3A. Use either the Winegard F-175 and WH-1 weatherproof housing or the weatherproof Finco G-514.

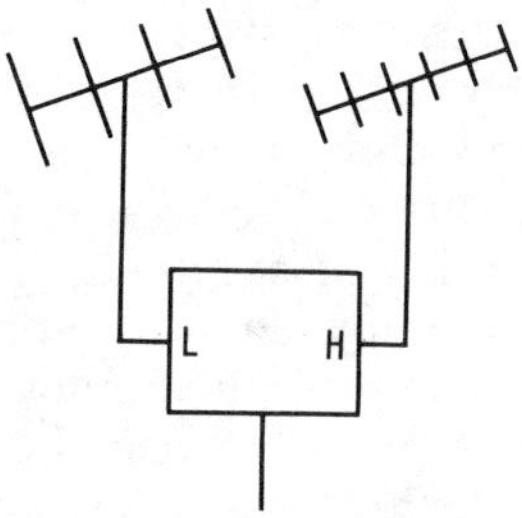

(A) Vhf-L and vhf-H.

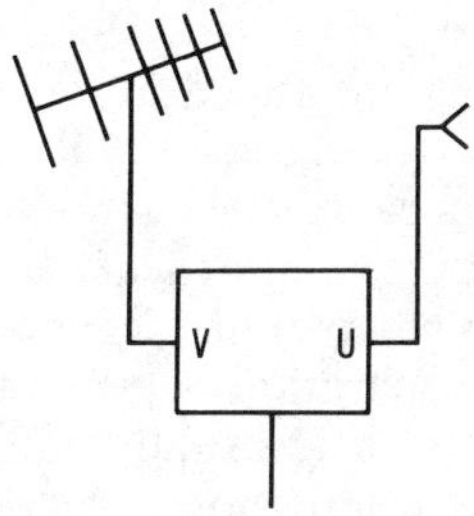

(B) Vhf and uhf.

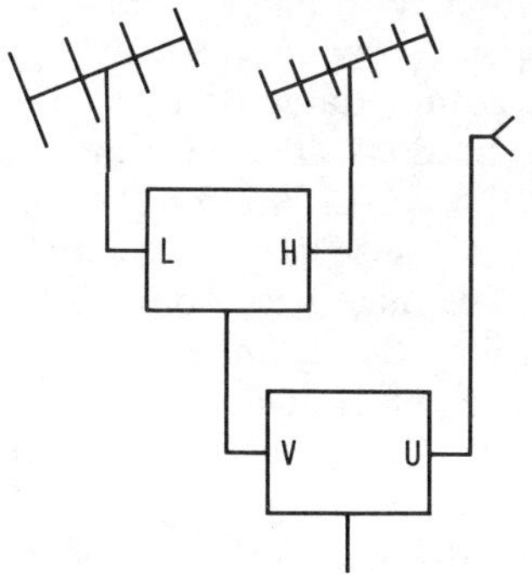

(C) Vhf-L, vhf-H, and uhf.

Fig. 8-3. Combining techniques for different-band antennas.

2. To combine any vhf antenna and a uhf antenna, use a 75-ohm vhf-uhf band separator backwards as a combiner, as shown in Fig. 8-3B. Use either the Winegard CS-775 and WH-1 weatherproof housing or the weatherproof Finco G-522.
3. To combine a vhf-L, a vhf-H, and uhf antenna, use the technique shown in Fig. 8-3C. First combine the vhf antennas as

described in Rule 1 above. Then treat the output of the vhf H/L band combiner as a vhf antenna, and combine it with the uhf antenna as described in Rule 2.

Any metallic object a half-wavelength or longer at its operating frequency can affect the operation of an antenna. In practice this means that the reflector rods of a uhf antenna can affect the high-band operation of a vhf antenna, a vhf low-band antenna can affect a vhf high-band antenna, etc. Because of this, antennas mounted on the same mast must be spaced far enough apart so their effect on each other is minimal. The amount of spacing depends a lot on the size and orientation of the antennas. A Channel 12 antenna with an 8-foot boom will have little effect on a Channel 2 antenna fairly close if they are both pointing in the same direction, because the Channel 12 antenna's elements are too small (a fraction of a wavelength at Channel 2) to have significant effect. However, if the stations these antennas are aimed at lie 90° (or nearly so) apart, the *boom* of the Channel 12 antenna will disrupt the operation of the Channel 2 antenna because the 8-foot boom is a half-wavelength at Channel 2! For this reason, at least 8 feet of separation must be maintained between vhf antennas when one of them is a low-band antenna. Three feet of separation between the closest portions of a uhf antenna and a vhf antenna is usually sufficient; the exception is when the uhf antenna is a long-boom yagi (like the 79-inch Jerrold CYD antennas) and it is set crosswise to a low-band vhf antenna. Then maximum separation (over 8 feet) is required.

The spacing requirements mean that no more than two antennas can be accommodated on masts that are practical for home and small matv installations. The best approach when three antennas are used is to put the vhf-L and vhf-H antennas on separate masts at opposite ends of the house (or at least 15 feet apart). Put the uhf antenna on the vhf-L mast with about 3 feet of separation between antennas. If it is a long-boom antenna mounted crosswise to the vhf antenna, put it on the vhf-H mast.

When two antennas are placed on the same mast, the question arises of which one goes on top. Since the topmost position is generally the favored position for reception, you might want to put either the antenna receiving the *weakest* station or the antenna receiving the most *important* station at the top. If mechanical stability of the antenna system is important (such as in windy areas), it is more sensible to put the smallest antenna on top. Another criterion is sheer convenience; if a uhf antenna with a vhf feedthrough provision is used, mounting the uhf antenna below the vhf antenna results in the shortest and neatest transmission-line connections (Fig. 8-4). If the choice is between a vhf-L and a vhf-H antenna, the vhf-L an-

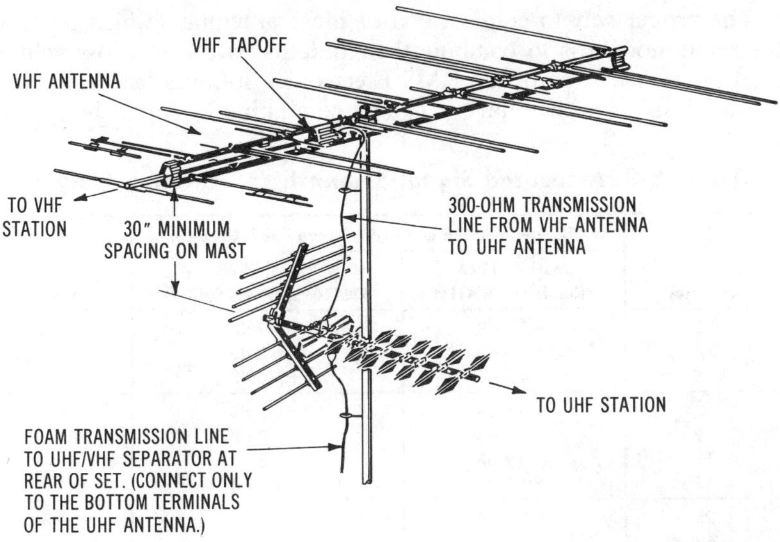

Courtesy Channel Master

Fig. 8-4. Vhf-uhf antenna installation using uhf antenna with vhf feedthrough.

tenna should be top mounted for the reasons discussed in "Higher Antennas."

Stacked Antennas

Stacking is a method of obtaining higher gain by combining the outputs of two or more identical antennas. Each time the number of identical antennas is doubled, the theoretical gain is increased by another 3 dB. Using two antennas instead of one can increase gain by up to 3 dB; using four will increase gain by up to 6 dB. Exactly how much gain increase will result depends on several factors, including the losses in the combining network, the impedance matching throughout the system, and the effective illumination of the antenna elements.

There are two ways to combine the individual antenna outputs. The simplest way is by mechanically connecting the antennas in parallel via stacking kits that are little more than aluminum rods simulating a transmission line. Few of the kits provide the proper impedance transformation needed between each antenna and the common feed point. The resulting mismatch limits the achievable gain to about 2 dB. None of the kits provides isolation between the two antennas, so if one is not sufficiently illuminated, it acts as a load for the well-illuminated antenna and diverts signal power from the tv set. This makes it possible to end up worse off than before.

The proper way to connect two or more antennas (which must be the same model) is to combine their outputs with a low-loss splitter used in reverse. One of the RMS Electronics splitters listed in Table 6-3 is ideal, since they have very low combining loss and their small

Table 8-3. Measured Signal Strength vs. Antenna Height

Channel	Change in Antenna Height Above Roof (Ground) Mark-8, 1977 8(33) ft → 16(41) ft	DJR-6, 1978 9½(34½) ft → 18(43) ft	Band
2	+1.6 dB	+5.1 dB	VHF-L
4	+5.0 dB	+6.0 dB	
5	+4.7 dB	+5.0 dB	
7	+0.9 dB	+0.3 dB	VHF-H
9	+0.4 dB	0 dB	
11	−0.9 dB	−0.3 dB	
13	+1.3 dB	−1.1 dB	
25	+3.5 dB		UHF
68	+5.5 dB		

size and waterproof construction are well suited to mast mounting. The excellent isolation of these splitters ensures that nearly the full output of each antenna will reach the transmission line regardless of the relative illumination of the antennas.

Although most antennas are basically 300 ohms output impedance, their outputs must be converted to 75 ohms to perform the combining. This is because even the best 300-ohm splitters have relatively high losses, and the combining gain is too low at uhf frequencies to warrant the cost and effort of a multiple-antenna installation. So, connect a low-loss outdoor balun (see "Baluns" in Chapter 6 for recommended models) to each antenna to be stacked. Be certain to connect each balun in the same way (same side up, same approach to the antenna) so the antenna outputs will be phased properly. Then connect the baluns to the "output" terminals of the appropriate splitter with equal lengths of identical coax. At uhf frequencies the lengths of the cables must match within ¼ inch for best results; at vhf frequencies they can be off by as much as 1 inch without trouble developing.

For best results, a field-strength meter should be used to check the phasing. Measure the output of the stack with the baluns arranged as described in the previous paragraph, and record the signal for each channel. Then reverse the connection to the antenna terminals of *one* balun, and remeasure the signal levels. If you did the job right initially, all (or at least *most*) of the signal levels will be lower. If they are, restore the balun to its original connection. If the newly measured levels are all higher than the original levels, it means the baluns were originally incorrectly phased, and the present connections should be retained. If you don't have a field-strength meter, you can make a crude but usable check with an attenuator and tv set. Add sufficient attenuation between the stack output and the tv until the picture becomes snowy. Then reverse one balun and see if the picture gets worse (as expected) or better.

As far as gain is concerned, there are two ways to stack a pair of antennas: vertically and horizontally. Vertical stacking is much easier since both antennas mount on the same mast, but on the vhf low band the amount of mast space required for proper spacing is a problem. Horizontal (side-by-side) stacking eliminates this problem, and ensures equal illumination when the signal level changes greatly with height. However, it is more difficult to accomplish mechanically, particularly with vhf antennas. With either method the spacing requirements are such that stacking is not recommended for broadband (12-channel) vhf antennas or vhf-uhf antennas. Best results are obtained from stacking single-channel antennas, and fairly good results are obtained from stacking single-band (vhf-L, vhf-H, or uhf) antennas. Refer to the section "Ghost and Interference Elimination" later in this chapter for details on each stacking method, and the unique alterations of polar patterns they produce.

Preamplifiers

Since each time you double the size of an antenna array you pick up 3 dB gain, simple arithmetic shows that you need a stack of eight antennas to add 9 dB gain to your system. This is impractical in most circumstances, so the alternative when high gain is needed is to use a preamplifier. A good preamplifier can provide 15 to 24 dB gain, an amount that would require 32 to 256 antennas to achieve!

The choice of a preamplifier is extremely critical. The preamplifier must have a noise figure lower than that of the tv receiver or distribution amplifier it drives if the preamplifier is to produce any benefits other than overcoming downlead loss. As detailed in Chapter 6, broadband preamplifiers with noise figures as low as 2.2 dB at uhf frequencies and 3 dB at vhf frequencies are available. Preamplifiers with high noise figures are useless with very weak signals; all you will get is a high-level noisy signal.

A mast-mounted preamplifier has the advantage that the signal is amplified before it undergoes downlead losses. This gives the best picture signal-to-noise ratio for two reasons: the signal-to-noise ratio is improved (in decibels) by the amount of the downlead loss (in decibels), and by the decibel difference in the noise figure of the preamplifier and that of the tv set or distribution amplifier. For example, suppose the vhf signal delivered to the tv set produces a

Grade B picture (34-dB signal-to-noise ratio). Installing a mast-mounted preamplifier with a 3-dB noise figure improves the picture signal-to-noise ratio by 4 dB because of the difference between the 7-dB noise figure of the vhf tuner and the 3-dB noise figure of the preamplifier, and by another 4 dB because of the loss of 100 feet of cable at 200 MHz. The 8-dB total improvement yields a picture signal-to-noise ratio of 42 dB, almost that of a Grade A picture. On uhf frequencies the improvement is more spectacular: 11 dB improvement due to the difference between the 13-dB noise figure of the typical tuner and the 2 dB of a super-low-noise preamplifier, and 6-dB improvement because of 100 feet of RG6U cable loss at 800 MHz.

The disadvantages of preamplifiers are twofold. First, preamplifiers do add some noise to the signal, so gain produced by preamplifiers is not as beneficial as the *same amount* of gain produced by more antennas. However, since the load of the antenna system is usually a high-noise device (such as the uhf tuner in a tv set), the relatively small amount of noise contributed by a low-noise preamplifier is insignificant, and the net result is a great improvement in picture signal-to-noise ratio (due to the reasons described above). The second problem is cross modulation. Because they contain active devices, preamplifiers can be overloaded by strong signals. Even if overload does not occur, a local signal can impress its modulation on a very weak distant station (cross modulation) if the local signal is strong enough.

GHOST AND INTERFERENCE ELIMINATION

Interference to tv reception can come from many sources, including the desired station (ghosts). The methods of dealing with interference are also many; some techniques are effective against many forms of interference, others are selective. The interference-elimination techniques that are effective against all forms of interference (ghosts, CB harmonics, motor noise, etc.) are antenna techniques. They utilize the directionality of an antenna or antenna array to pick up only signals coming from the desired direction, and reject those coming from other directions. The selective-elimination techniques pass or reject signals picked up by the antenna on the basis of frequency; devices that do this are called *filters*.

The first thing to do when ghosts or interference symptoms appear in the picture is to determine which direction (*relative to the direction of the desired tv station*) the interference is coming from. When this is known, the course of action can be determined. For example, Fig. 8-5 shows an omnidirectional tv transmitter (T) and the paths of both the direct signal to a tv receiver and several reflected (ghost) signals. If the interference is coming from behind, as is the reflection from water tank WT, the cure is fairly easy: an antenna with very high front-to-back ratio. Extremely persistent cases may require stagger stacking. If the interference is coming from the side (apartment building AB), an antenna with deep side nulls should be selected. However, if the angle between the interfering signal and the desired (direct) signal is very small (less than 65° for low-band vhf, 35° for high-band vhf, or 40°–60° for uhf), no single antenna suitable for home or small matv installation will be of much use. An array consisting of horizontally stacked identical antennas is required in this case to handle ghosts (like the one coming from tall building TB in Fig. 8-5) or on-frequency rf interference. If the interference is off-frequency, a filter can be used.

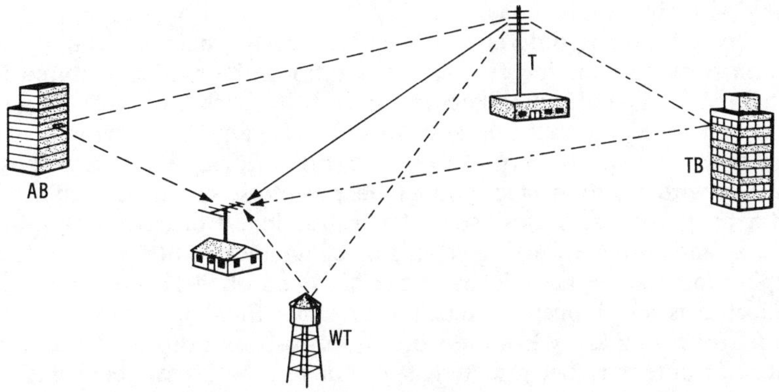

Fig. 8-5. Direct and reflected signals reaching a tv antenna.

High-Performance Antennas

High performance antennas are those that come closest to approximating qualities of the ideal directional antenna: zero pickup from the back and sides. The front-to-back ratio tells how close a real antenna comes to rejecting signals from a direction 180° opposite to the forward direction, while an inspection of the horizontal-plane polar pattern is necessary to determine how well it will reject signals coming from the sides.

Small, low-gain antennas *generally* have poor directionality, so they have a considerable back lobe. Moreover, simple all-channel vhf antennas and vhf-uhf antennas have sizable minor (side) lobes because of multimoding. As the number of elements on these types of antennas increases, the side and back lobes shrink (provided the designer is doing a good job!). The larger-size antennas, particularly those with relatively long booms, also have good vertical-plane directionality; in practice this manifests itself in good nulls in the horizontal-plane pattern. An antenna with these characteristics (high front-to-back ratio, small side lobes, and deep nulls) is the primary method of eliminating ghost pickup from the sides and back. Since these characteristics are found (with one exception) only on large high-gain antennas, a large antenna is required for ghost elimination even if the signal is quite strong. The increased signal output obtained from switching to a large antenna can be attenuated if necessary. The one exception is a type of antenna designed for high directionality at the expense of gain. The Winegard CH-4210 described in Chapter 2 is a good example of just such an antenna, although it cannot match the performance of very large (over 10 feet long) conventional antennas. Either way, pickup of signals from the sides and rear, be they ghosts or rf interference, is much less than with simple antennas.

The situation is different with uhf antennas; there is little correlation between gain and front-to-back ratio in the models featured in Chapter 3. The deciding factor for front-to-back ratio of uhf antennas is essentially the quality of the reflector. For example, the antennas with the best front-to-back ratios in Chapter 3 were those having wire-mesh or steel-grill plane reflectors, or corner reflectors with apex rods and close-spaced construction. Corner-reflector antennas and corner-reflecting yagis missing an apex rod (typical of their type) have sizable back lobes at the high end of the band where the reflector is no longer sufficiently opaque. Finally, the very-high-gain Hoverman array has the worst front-to-back ratio of the recommended antennas because it has just a few half-wave elements to act as a reflector for a large area of driven element. While an antenna of this type is excellent for fringe areas (where ghosts *rarely* are a problem), it is a poor choice for a metropolitan area (where ghosts *usually* are a problem). The best ghost-fighting antennas for uhf frequencies are of the 4-bay bowties or the RCA corner reflectors. These antennas also have good vertical-plane discrimination, so they are better than most others for rejecting *ground* and aircraft reflections.

The statements made for vhf antennas also apply to the vhf sections of vhf-uhf antennas. Similarly, the statements about uhf corner reflectors also apply, only the situation is worse because the corner reflector of vhf-uhf antennas tends to be small and always has a wide gap at the apex to permit passage of the feeder lines from the vhf section. Larger vhf-uhf antennas have bigger corner reflectors, which helps somewhat, but they also have vhf directors out front on the boom ahead of the uhf-driven element. These vhf elements act as uhf reflectors at some frequencies and allow considerable pickup from the rear. This is another reason why the author does not recommend the use of large vhf-uhf antennas. When high performance

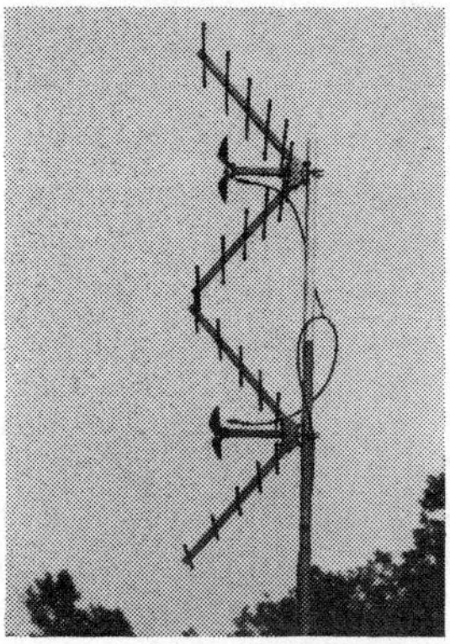

Fig. 8-6. Vertically stacked uhf antennas.

is needed for both uhf and vhf reception, separate uhf and vhf antennas should be used and combined with a high-performance vhf-uhf band separator/combiner, as described earlier.

Vertical Stacking

In "Increased Signal Output" you learned how stacking antennas can increase gain, and details of how to properly connect two or more antennas. However, the greatest reason for stacking antennas is for the dramatic alteration in polar patterns that occurs.

Stacking antennas vertically (Fig. 8-6) increases gain but provides no significant change in directivity in the horizontal plane. However, a spectacular improvement in the vertical-plane pattern does result (Fig. 8-7). This can be extremely helpful in eliminating ghosts caused by ground or roof reflections, airplane flutter, and electrical noise from any dwelling near but below the antenna. Most kinds of antenna can be stacked vertically, the most important consideration being the distance between the two antennas. To secure close to the maximum-possible gain increase and a desirable vertical-plane polar pattern, the antennas must be properly spaced. They must be separated enough so each does not intrude on the effective

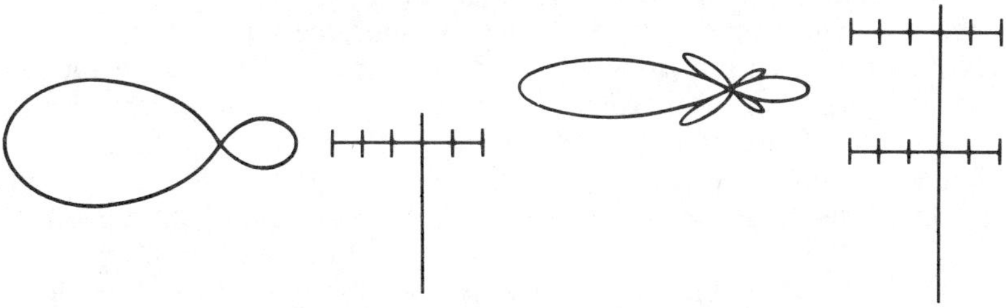

Fig. 8-7. Change in vertical-plane polar pattern caused by stacking.

aperture of the other, although stacking them too far apart (which is likely only at uhf frequencies) makes equal illumination difficult under some circumstances and results in vertical-plane pattern peculiarities (excessive minor lobes). Spacings commonly are 0.6 to 1.0 wavelength. The optimum spacing depends somewhat on antenna gain; high-gain antennas require greater spacing than low-gain antennas. Spacings less than one-half wavelength are rarely used, as the resulting gain increase and pattern improvement are not worth the effort or expense.

Vertically stacking vhf antennas is complicated by several factors. First, broadband vhf antennas cover a 4:1 range. This means that the spacing yielding greatest gain improvement on Channel 2 will be too large for Channel 13, and the spacing yielding greatest gain improvement on Channel 13 will be too small to show improvement on Channel 2. For this reason broadband vhf antennas are rarely stacked by the knowledgeable antenna installer. However, it is often possible to get good results with single-band antennas, high-band vhf antennas in particular. Because of the mere 1.2:1 frequency ratio of the vhf high-band channels, a spacing optimized for Channel 10 will give very good results on all high-band channels. Because of their greater (1.6:1) frequency ratio, low-band antennas are usually spaced at a distance that is about 0.6 wavelength on Channel 2 and one wavelength on Channel 6. However, the biggest problem in stacking low-band antennas, be they single-band (Channels 2–6) or single-channel antennas (where no compromise in spacing is necessary), is the matter of the absolute distances involved. A half wavelength on Channel 2 is over 8 feet. This means that the antennas alone occupy most of a 10-foot mast section, making rotation difficult if not impossible. Furthermore, *at least* another 10-foot mast section is required to hold the bottom antenna a decent distance from the roof. Twenty feet of masting requires guying, thus limiting the choice of mounts. For this reason, vertically stacked low-band antennas (be they single-channel or single-band) are found only in large matv and catv installations, where tall towers raise the antennas to a height allowing fairly equal illumination, and rotation is not a factor. Table 8-4 gives the recommended spacings for single-channel and single-band vhf antennas. The single-channel data represents the recommendation of several manufacturers of single-channel yagis. Since some manufacturers (Jerrold and Winegard) favor wide spacing, and others (Finco and SITCO) favor minimum spacing, a distance range is given for each channel. The shorter distances represent about a half-wavelength spacing, the longer distances about 0.65 wavelength on Channels 2–6, and 0.8 wavelength on Channels 7–13. Obviously, practical factors (i.e., available mast space) have influenced their selections. As such, the shorter distances should be used only for medium-gain

Table 8-4. Vertical Stacking Distances for VHF Antennas

Channel(s)	2	3	4	5	6	7	8	9	10	11	12	13	2–6	7–13
Min. Distance (in)	103	94	86	75	70	33	32	31	30	29	29	28		
													132	54
Max. Distance (in)	140	127	116	102	94	54	52	50	48	46	45	44		

antennas (five elements or so). The wider spacing is mandatory for high-gain antennas (ten elements or so), and recommended for all. In fact, if gain (rather than pattern improvement) is important, a spacing around 0.9 wavelength should be used with all high-gain antennas.

The recommendations for single-band antennas yield 0.6 to 1.0 wavelength spacing on the vhf low-band (Channels 2–6), and 0.8 to 1.0 wavelength spacing on the high band (Channels 7–13).

Finding sufficient mast space is rarely a factor in vertically stacking uhf antennas. The problem is usually one of spacing them close enough, since the vertical dimensions of uhf antenna types using reflector screens determine the minimum-possible spacing. For instance, one wavelength on Channel 14 is 25 inches. The reflector on all corner reflectors and most corner-reflector yagis have greater height than this, so the resulting center-to-center spacing for the array will be greater than 25 inches. Therefore, uhf corner reflectors (like the RCA 7B140s shown in Fig. 8-6) should be stacked as close as their reflector screens permit. The same applies to corner-reflector yagis. A pair of Antennacraft G-1483s (an excellent far-fringe-area array), however, should be spaced about 6 inches apart so the tuned reflector elements will work properly. About the only uhf antenna types that can be closely spaced are yagis using a single tuned reflector element or small screen. However, most of these (like the single bowtie reflector antenna) are low-gain antennas which should be replaced by a high-gain version (big corner-reflector yagi or 4-bay bowtie) instead of being stacked.

Stagger Stacking

Stagger stacking is a special variation of vertical stacking in which one antenna is mounted a quarter wavelength ahead of the other (Fig. 8-8). The cable section connecting the forward antenna to the signal combiner is made an *electrical* quarter wavelength longer than the other so the signals from the front are combined in phase. Signals arriving from the back, however, are 180° out of phase and thus cancel. This results in very high front-to-back ratio, even if the front-to-back ratio of the antennas alone is poor. When this technique is used on antennas having a high front-to-back ratio to begin with, the rearward pickup of the array is virtually zero. Unfortunately, this

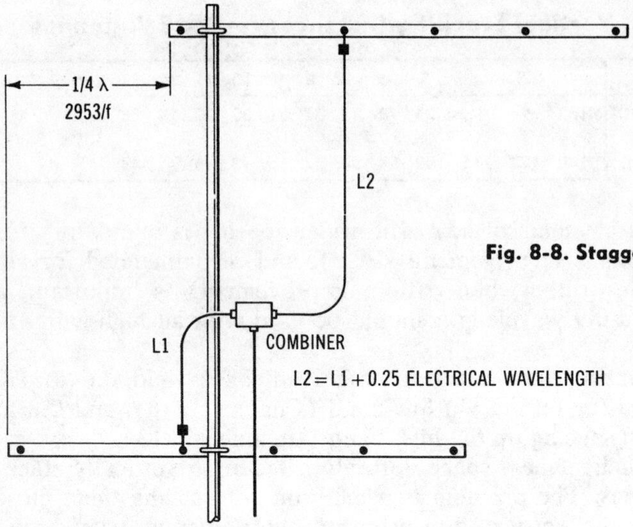

Fig. 8-8. Stagger stacking.

technique is frequency sensitive; it works best at one frequency, and is useful only over a relatively small frequency range (±10 percent). Therefore, it should only be used on single-channel or vhf-H single-band antennas. It is also difficult to accomplish mechanically; the mounting clamp of the staggered antenna must be moved back on the boom a quarter wavelength, and a support built to hold up the front end.

Its main application is where an extremely strong ghost (nearly the same strength as the direct signal) comes from behind, or where on-frequency rf interference (a CB harmonic, for example) originates behind the antenna. Stagger stacking is especially suited to this last application because CB harmonic interference is often very severe on one channel. In cases like this, a pair of stagger-stacked single-channel antennas can be erected for the problem channel, and the output of this array can be combined with that of the other antenna(s) by means of a yagi coupler or one of the other techniques shown in "MATV HEAD ENDS."

Table 8-5 shows worked-out dimensions for CB harmonics as an example of how to calculate the amount of stagger (displacement) and the additional length of transmission line needed. Notice two things. First, an electrical quarter wavelength of transmission line is shorter than a quarter wavelength in air (antenna displacement). You must multiply the air dimension by the *velocity factor* of the transmission line. The velocity factor for all foam coax is 0.78; for standard twinlead and Belden 8275 it is 0.8. Second, the electrical quarter wavelengths are the *difference* in length between the two line sections connecting the antennas to the signal combiner.

Table 8-5. Staggering Dimensions

Rejection Frequency (MHz)	Channel	Antenna Displacement (in)	Electrical ¼ Wavelength Foam Coax (in)	Electrical ¼ Wavelength Standard Twinlead (in)
54	2	54¾	42⅝	43¾
81	5	36½	28½	29¼
189	9	15⅝	12⅛	12½

Horizontal Stacking

Horizontal stacking is accomplished by mounting two identical antennas side by side, and combining their outputs. Aside from providing up to 3 dB more gain than one antenna alone, this arrangement produces an array horizontal-plane polar pattern having a narrow main lobe with extremely deep nulls along the sides (Fig. 8-9). This is the ideal pattern for eliminating ghosts arriving from a direction near that of the desired signal.

The beamwidth of the main lobe and the angular distance of the nulls depends on the spacing (in terms of wavelength) between the two antennas. Center-to-center spacings of 0.6 to 1.2 wavelengths are most commonly used, although much greater spacing is used in special circumstances. The wider the spacing, the narrower the beamwidth and null angles. However, the size of the side lobes (minor forward lobes adjacent to the main lobe) also increases with the spacing. This is evident from the example shown in Fig. 8-9, where relatively wide spacing (1.2 wavelengths) was used.

A pair of wide beamwidth antennas can provide an array with a narrow-beamwidth main lobe because of an effect called *pattern*

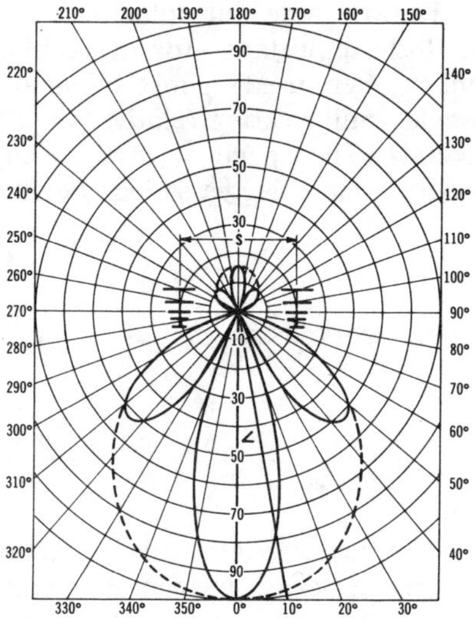

Fig. 8-9. Change in horizontal-plane polar pattern produced by horizontal stacking.

multiplication. In fact, with the aid of a trigonometric calculator, you can determine the effects of horizontal stacking by multiplying the relative field strengths of the horizontal-plane polar pattern (such as those appearing in this book) of any antenna by the appropriate array factors, and replotting the new array polar pattern. The formula for calculating the multiplier (array factor, af) at each bearing is given below. As an example of how to use it, look at the 10° bearing in Fig. 8-9. The relative field strength (rfs) of the basic antennas at this bearing is 98. With 1.2 wavelengths (432°) spacing, the multiplier for this bearing is 0.793, yielding a relative field strength for the array of 77.7 at the 10° bearing.

Given:

$$rfs_{array} = rfs_{ant}(af)$$

$$af = \cos\left(\frac{s}{2}\sin \angle\right)$$

$$s = 432°$$

$$\angle = 10°$$

$$rfs_{ant} = 98$$

Finding Array Relative Field Strength:

$$af = \cos\left(\frac{432°}{2}\sin 10°\right)$$

$$= \cos\ (216° \times .1736)$$

$$= \cos\ (37.5°) = 0.793$$

$$rfs_{array} = 98\ (0.793) = 77.7$$

Knowing the full polar patterns of a horizontally stacked array is of little importance for the application where horizontal stacking is most often used: ghost elimination. For this application the null angle (null bearing relative to that of the main lobe) is what really counts. The spacing (in wavelengths) required for a specific null angle $\angle$, and vice versa, can be determined by the formulas

$$S_\lambda = \frac{1}{2\sin \angle} \quad \text{and} \quad \angle = \arcsin \frac{1}{2S_\lambda}$$

Some worked-out values of null angle vs. spacing are also given below.

Spacing	0.6	0.8	1.0	1.2	1.5	2.0	Wavelengths
Null Angle	57	39	30	25	20	15	Degrees

The field strength in any direction, *relative to the main lobe,* can be no greater after horizontal stacking than it is on the basic antenna.

The side lobes appear because relative gain was *reduced* in the null bearings, not because the side-lobe bearing gain has increased relative to the forward direction. For this reason, if the basic antennas themselves have narrow beamwidth (as is the case with ten-element single-channel antennas), the array polar pattern will be lobe-free even with modestly wide spacing. A side lobe cannot "appear" if the basic antennas have little or no pickup in that direction. This is shown by comparing Fig. 8-10 to Fig. 8-9. The spacing between antennas is the same (1.2 wavelengths), but the basic antennas have a significantly narrower beamwidth (57° vs. 76°). Although the resulting main lobe is *very slightly* narrower, the big difference is in the size of the side lobes. If extremely narrow beamwidth antennas (like the SITCO CA 12-1) are used, the array pattern has no side lobes at all.

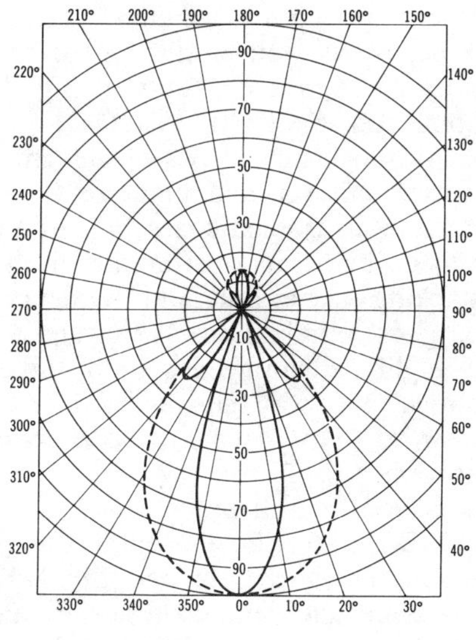

Fig. 8-10. Small side lobes resulting from use of narrow-beamwidth antennas.

The adjustment of spacing to null out a ghost can only be done for one channel for each pair of antennas. In theory, 12-channel vhf antennas can be used for ghost elimination on one channel, and simply as a high-gain array on the others. In practice, it rarely works out. For example, if the ghost is on a high channel, the spacing distance needed to null out the ghost might be less than the length of the antenna elements. Similarly, if the ghost is on a low channel, the spacing distance might be so great that the beamwidth on the high channels is too narrow for stable reception. A beamwidth of less than 10° is too critical to set with a rotator, and it will also cause rapid signal fluctuations as winds shake the antenna. For these reasons horizontal stacking at vhf frequencies is done primarily with single-channel and high-band antennas.

Horizontally stacking vhf (and some uhf) antennas is much harder to accomplish mechanically than vertically stacking them. This is because two short masts are needed on which to mount the antennas. These must be attached to a crosspiece which is in turn attached to the main mast or tower. A mast section or length of metal pipe cannot be used for the crosspiece since it is parallel to the antenna elements and anything metallic will adversely affect the operation of the antennas. Either a redwood beam or a framework of plastic pipe is commonly used for the crosspiece. Moreover, the coaxial cables connecting the antennas to the combiner cannot be run horizontally either. They should be run vertically as much as possible and afterwards at a 45° angle. Another problem peculiar to vhf low-band antennas is the absolute dimensions of the crosspiece. One wavelength at Channel 2 is about 17 feet, and 17 feet of 2×4 is heavy, as well as difficult to acquire. For this reason, horizontal stacking is practical only at high-band vhf and uhf frequencies for home and small matv systems.

Although most types of uhf antennas can be horizontally stacked, corner-reflector and plane-reflector antennas are easiest. Since the horizontal crossbars are behind the reflector screen, they have no effect on antenna operation even though metal. This makes horizontal stacking easy to do at uhf frequencies, whether it be for increased gain or ghost elimination. When ghost elimination is most important, it is highly advantageous to be able to vary the spacing of the antennas. This way the angular difference between the desired signal and forward ghost need not be precisely known beforehand; the entire array is rotated to peak the desired signal, then the antenna spacing is varied to null out the ghost. The RCA 7B140 (Figs. 8-11A to 8-11D) is ideal for this application, since it can be readily modified to mount on a horizontal bar (Fig. 8-11C). Drill four holes in the mounting plate and antenna bracket, then bolt the plate to the antenna bracket at a right angle to its normal orientation (Fig. 8-11D). However, if the desired spacing is known, or minimum spacing is desired, adjustability is not needed. In this case, a much lighter array will result if the mounting plate and fittings from each 7B140 are discarded and the antenna simply bolted to aluminum tubing or channel with square cross section.

Plane-reflector antennas (bowtie-reflector and 4-bay bowties) must be bolted to holes in the cross member(s), so they are practical only when the correct spacing is known or just a high-gain narrow-beam antenna is desired. Aluminum tubing with square cross section is ideal for the cross members. The patterns of a pair of old-style JFD UHF-202s (Fig. 8-12A) spaced 17½ inches between centers are

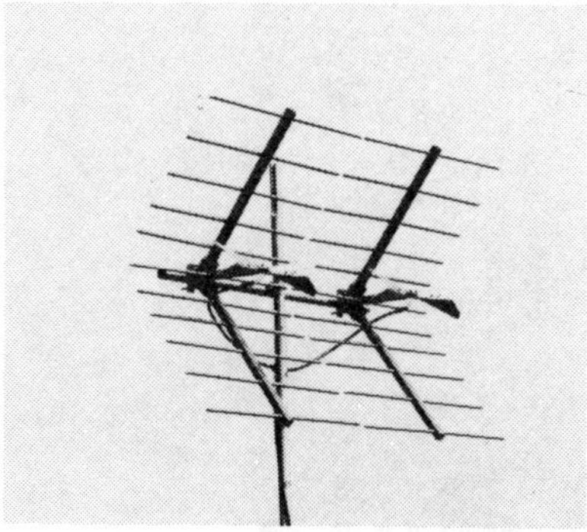

Courtesy Sony Corp. of America

(*A*) *Horizontally stacked corner-reflector antennas.*

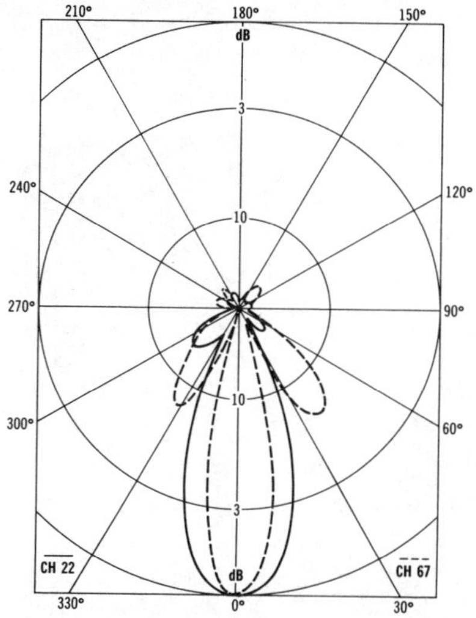

(*B*) *Polar patterns of horizontally stacked corner-reflector antennas.*

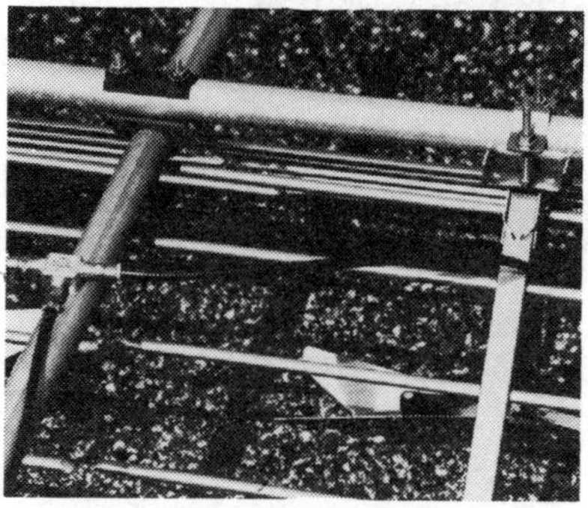

Courtesy Sony Corp. of America

(*C*) *Closeup of horizontal stacking bar.*

(*D*) *Modified 7B140 mounting bracket.*

Fig. 8-11. Horizontal stacking of the RCA 7B140 antennas.

shown in Fig. 8-12B. Note the small side lobes resulting from the relatively close spacing.

Commercially built arrays using horizontally stacked antenna elements are available for vhf and uhf reception. Nearly all of those made for vhf service are ruggedly built single-channel arrays intended for catv and large matv systems. Very often they are part of *quad stacks,* like the SITCO model CA-48-4 shown in Fig. 8-13. This is obviously too large and expensive an array for home use or for

Courtesy Sony Corp. of America

(A) Stacked 4-bay bowtie antennas.

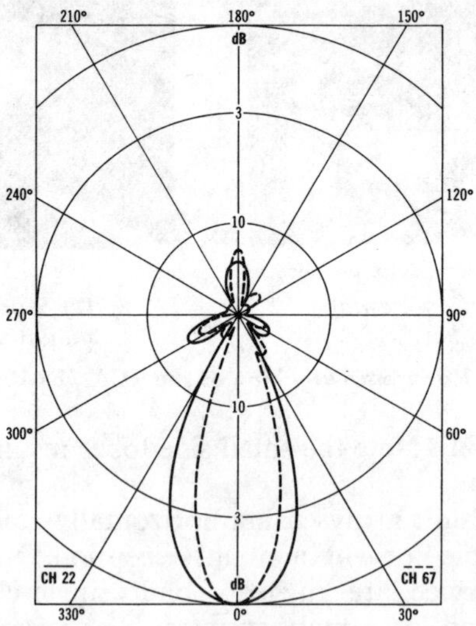

(B) Polar patterns oj antennas in (A).

Fig. 8-12. Horizontally stacked plane-reflector antennas.

a small matv system, but the situation is different at uhf frequencies. Several manufacturers sell antennas that are equivalent to side-by-side 4-bay bowties. They are usually called *8-bays.* Some are actually a pair of 4-bays with mounting brackets and stacking harness, similar to what the author did in Fig. 8-12. Others are actually integrated antennas, featuring a single reflector screen. A pair of 4-bay bowtie driven elements are mounted side by side against a continuous wire-mesh reflector, and connected in parallel. However, to realize the

(*A*) *Single-channel quad stack.*

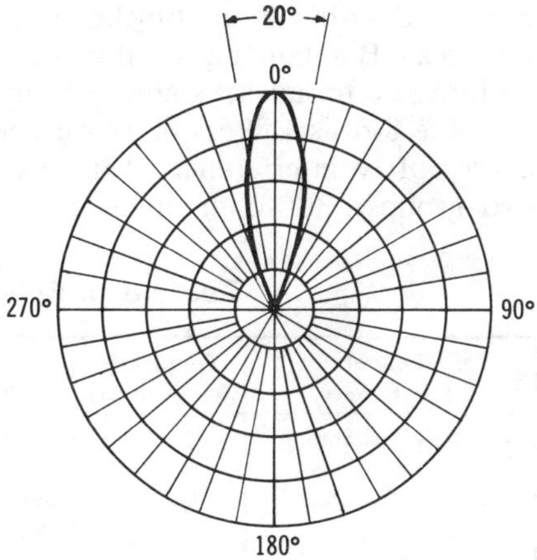

(*B*) *Horizontal-plane polar pattern.*

Courtesy SITCO Antennas

Fig. 8-13. SITCO single-channel antenna and pattern.

maximum gain from this configuration, the individual wires forming the transmission-line sections connecting each 4-bay to the common output terminal must be precisely the same length. Furthermore, the spacing between the right and left driven elements must be wide enough to yield a reasonably narrow beamwidth at the low end of the uhf band, but close enough so the pattern at the high end (Channel 69) has tolerable side lobes. Not many 8-bays meet both of these requirements *and* have first-class mechanical construction, so you are often better off making your array from a pair of good 4-bays, and combining their output with a signal splitter.

Filters

When the interference is coming from the same direction as the desired tv signal, or antenna techniques are insufficient to do the job alone, filters are necessary. Rejection filters stop signals of a particular frequency, and let all others (including the desired tv signals) pass through the system. Although tv receivers contain selective circuitry designed to pass only the desired signal frequency, the preamplifiers and distribution amplifiers used ahead of the receivers are broadband devices that are more susceptible to overload caused by strong off-channel interference. The appropriate filter should be used whenever interference symptoms (such as cross modulation or wavy lines in the picture) appear in a tv antenna system using a preamplifier and/or distribution amplifier. An excellent way to determine what is causing the interference is to cruise the neighborhood or area and see what type of non-tv antennas are around. If a big ham antenna is located within a few blocks, it is a likely possibility. If a neighbor put up a CB antenna a few doors away, that is a possibility. If there is an fm radio broadcast station nearby, . . . well, you get the idea.

Table 8-6 lists filters of proven performance for each of the major causes of tv interference. The recommended filters are divided into two groups: 300-ohm outdoor and 75-ohm indoor. The 300-ohm

Table 8-6. Rejection Filters

Type of Interference	Frequency Range (MHz)	Filter Models 300-Ω Outdoor	75-Ω Indoor
FM Broadcast Stations	88–108	Winegard TP Series[1] Winegard TFM-3 Jerrold FMR-300 RCA 10G230	Winegard TFM-7 Jerrold FMR-75
Aviation Radio	108–135	Ch. Master 0211	Winegard BRF-170
Public Service Radio 2-Meter Ham Radio	144–160	Ch. Master 0211	Winegard BRF-170
6-Meter Ham Radio	50–54	Finco 3013	Drake TV75HP
TV Station	54–88 174–216 470–806	Winegard TP Series[1] Winegard UT-23[2]	Winegard TSC Series[1] Winegard UT-27[2]
CB Harmonics	54–54.4 81–81.6		Vitek VTV9M/MB Vitek VT7M/MB
CB Fundamental	27–27.4	Winegard TP-0270 Ch. Master 0227	Winegard HP-2700 A.C.A. MLX-405 RMS CA-2700 RMS CA-2600[3] Jerrold T-4040[3]
HF Ham Radio	1.8–29.4	Ch. Master 0227 Ch. Master 0211 Finco 3013	Winegard HP-2700 A.C.A. MLX-405 RMS CA-2700 RMS CA-2600[3] Jerrold T-4040[3]

[1] Selected to match interference frequency or channel.
[2] Tunable to specific channel.
[3] 75-Ω input, 300-Ω output to tv receiver terminals.

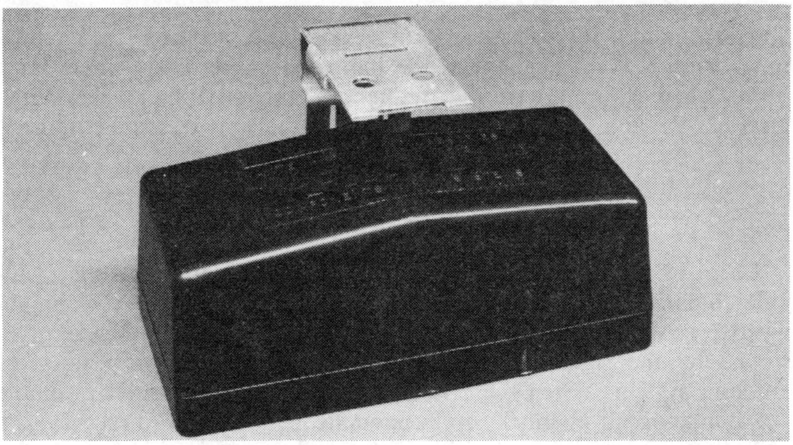

(*A*)*Photograph.*

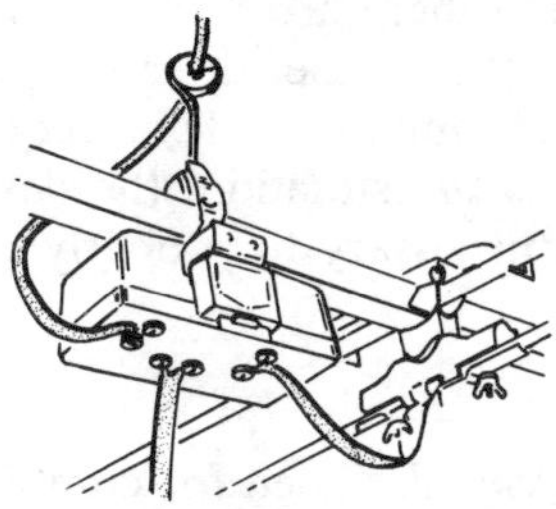

(*B*) *Boom.*

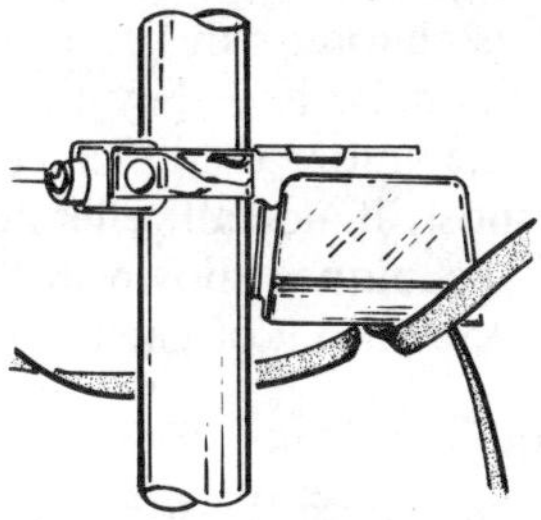

(*C*) *Mast.*

Courtesy Finney Co.

Fig. 8-14. Finco 3013 filter and mounting options.

filters are intended for insertion in the signal chain between the antenna and a mast-mounted preamplifier. Except for the Winegard TP-series filters, all 300-ohm models listed have weatherproof enclosures and mounting brackets that permit mounting on the antenna boom or mast. The Finco 3013 shown in Fig. 8-14 is a good example of this type of unit; it can be mounted almost anywhere, as the illustration shows. The Winegard TP filters are designed for installation inside their AC-series preamplifiers, which were recommended in Chapter 6.

The 75-ohm filters are intended for insertion in the signal chain between the antenna downlead and the distribution amplifier, if any. A few filters listed in the 75-ohm column are actually combination baluns and high-pass filters; they should replace baluns used to interface a 75-ohm signal-distribution system to the 300-ohm antenna terminals of a tv set.

Most of the rejection filters are fixed-tuned devices that stop the entire band indicated, or at least a big chunk of it. The Winegard

TP-series and TSC-series traps are devices that work on one channel only or on a small segment of the indicated band. These filters must be matched to the interfering frequency. Only two filters, the Winegard UT-23 and UT-27, must be tuned at the time of installation.

MATV HEAD ENDS

Large or small, matv systems must be able to supply all the available channels simultaneously, since the viewers at one receiver might wish to see a different channel than the viewers at another receiver served by the signal-distribution system. This means a rotatorless antenna system. When there is no more than one channel available per band, or all channels on a band lie in the same direction, the techniques described under the section "Separate Antennas" earlier in this chapter can be used. However, when there are a great number of channels, and they lie in different directions, the antenna system becomes more complicated. Single-channel antennas will be required for most, if not all, channels. When this is the situation, the devices and techniques shown in this section are necessary to properly combine the output of each antenna.

Yagi Couplers

A yagi coupler or signal injector is a device designed for combining single-channel (yagi) antennas, or for combining a single-channel and a broadband antenna. Yagi couplers contain both bandpass and bandstop filters tuned to a specific channel. The bandpass filter is connected between the yagi (CHANNEL) input and the coupler output; it *passes* only signals on the specified channel. The bandstop filter is connected between the broadband (ALL BAND) input and coupler output. It *stops* signals on the specified channel, but *passes* all signals on the other channels.

Fig. 8-15 shows how several single-channel yagi antennas are coupled. The number of yagi couplers needed is always one less than the number of antennas they must couple. These devices have 1.5 to 2.5 dB on-channel insertion loss at the yagi input but only 0.5 to 1 dB off-channel insertion loss at the broadband input. Because of this it is customary to order couplers matching the stronger channels (4 and 8 in the illustration), and connect the antenna for the weakest channel to the broadband input of the uppermost (Channel 8) coupler. The next antenna (Channel 8) connects to the yagi input of the uppermost (Channel 8) coupler. The output of this coupler (which now contains Channel 8 and Channel 13 signals) connects to the broadband input of the next (Channel 4) coupler. The Channel 4 antenna connects to the yagi input of the Channel 4

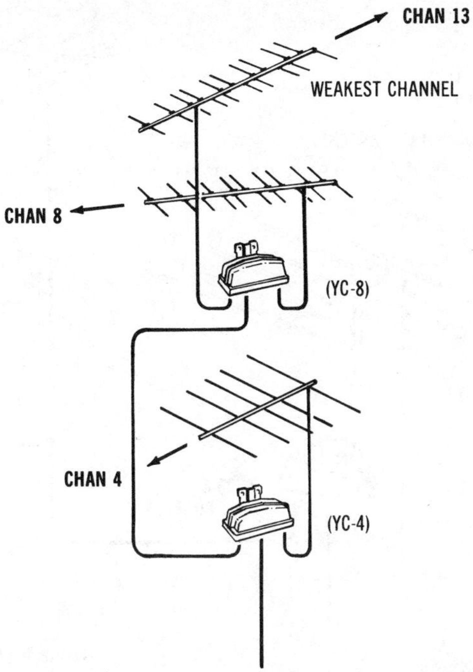

Courtesy TACO/Jerrold

Fig. 8-15. Combining single-channel yagi antennas with yagi couplers.

coupler, and the result is that the output of the Channel 4 coupler contains Channel 4, Channel 8, and Channel 13 signals.

Fig. 8-16 shows how a single-channel antenna or antenna array is coupled to a broadband antenna. This situation is likely to occur when one station lies in a different direction from the rest, or interference on one channel requires special antenna techniques (stagger stacking for example) for its elimination. The single-channel antenna or antenna array is connected to the yagi input of a matching (same channel) yagi coupler, and the broadband antenna is connected to the broadband input of the coupler. The bandpass and bandstop circuits in the yagi coupler block transmission of Channel 4 signals picked up by the broadband antenna, passing only Channels 2, 7, 9, 11, and 13 from this antenna, while passing only Channel 4 signals from the yagi.

Yagi couplers are available in 75 or 300 ohms impedance. Since they are always mounted near the antennas, they are equipped with weatherproof housings and mounting brackets. The Finco G-510 pictured in Fig. 8-17A is an exceptionally well made unit, having the cast metal housing, universal mounting bracket and fittings, and

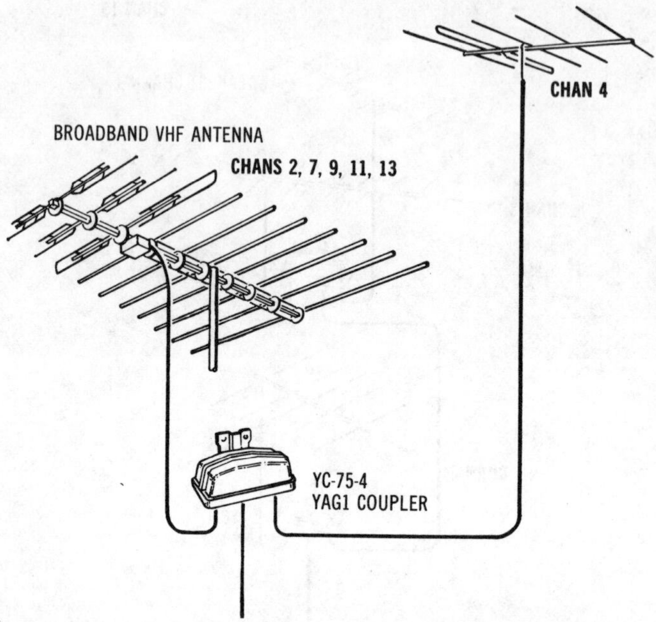

Courtesy TACO/Jerrold

Fig. 8-16. Combining a single-channel and a broadband antenna with a yagi coupler.

rubber boots for the cable connections that are standard on all of Finco's 75-ohm outdoor units. This unit has very flat bandstop characteristics and skirt selectivity in the ALL BAND signal path (see Table 8-7). The G-510 is therefore highly recommended for vhf antenna systems in severe climates.

The Channel Master 0573 Join-Tenna is a relatively-low-cost 75-ohm yagi coupler which also has good performance, particularly in regard to it being able to pass uhf signals (as from a vhf-uhf antenna) through its ALL CHANNEL port without loss. It also features unusually low (for a yagi coupler) on-channel attenuation at the yagi input.

The Channel Master 0273 Join-Tenna is an all 300-ohm version of the 0573. It is identical in performance and housed in the same type of plastic enclosure (Fig. 8-17B) as the 75-ohm version. Both of these units are highly recommended for antenna systems where uhf frequencies must be handled by the yagi coupler.

Channel-Mixing Networks

When all of the antennas on a band are single-channel antennas, and they number more than three, a channel-mixing network becomes

Table 8-7. Measured Performance of Yagi Couplers

Model	Impedance (Ω)	Port	Insertion Loss (dB) 50	60	63	66	80	200	550	Standing-Wave Ratio 50	60	63	66	80	200	550 (MHz)
Finco G-510-3	75	All Band	0.9	22	15	10	0.3	0.2	—	2.2	—	—	—	1.30	1.52	—
		Channel 3	29	2.4	1.2	2.4	32	66	—	—	1.93	1.70	2.0	—	—	—
Ch. Master 0573-3	75	All Channel	0.8	12	50	12	0.8	0.1	—0.1	2.5	—	—	—	1.85	1.03	1.25
		Channel 3	22	1.9	0.6	1.0	24	59	50	—	2.5	1.5	1.93	—	—	—
Ch. Master 0273-3	300	All Channel	0.9	10	31	14	0.4	0.0	0.0							
		Channel 3	23	1.6	0.7	1.2	25	50	40							

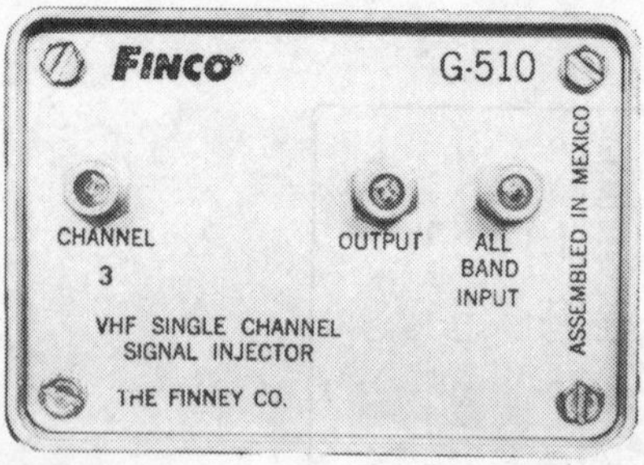

(*A*) *75-ohm.*

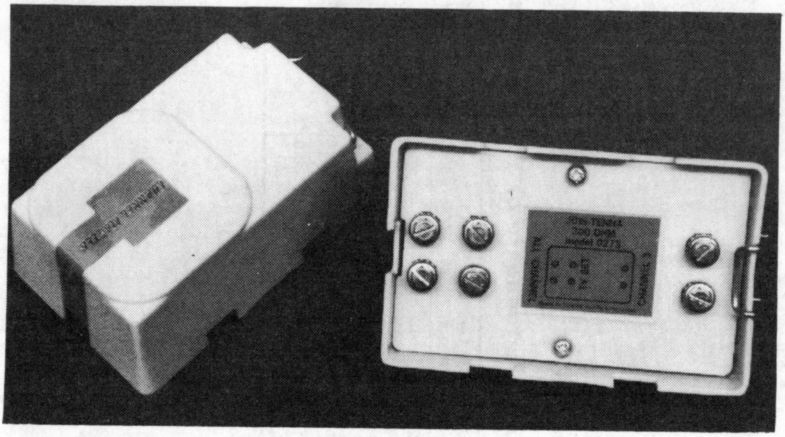

(*B*) *300-ohm.*

Fig. 8-17. Yagi couplers.

economical and convenient. These devices are generally available with five- to seven-channel capability. The vhf model shown in Fig. 8-18 can have any mix of low- or high-band channels, but adjacent channels are forbidden. (NOTE: Because of frequency separation, Channels 4 and 5, and 6 and 7, are not considered adjacent.) If both vhf and uhf channels are received, the outputs of the separate vhf and uhf channel-mixing networks must be combined with a band separator/combiner like the Winegard CS-775. The input connections to a channel-mixing network are straightforward; simply run coaxial cable from each single-channel antenna to the corresponding channel input on the mixing network. Barrel attenuators can be inserted between the antenna cables and channel inputs to attenuate very strong channels down to the level of the weaker ones if gain equalization is desired. This is shown for some of the channels in Fig. 8-19.

Courtesy Finney Co.

Fig. 8-18. Finco G-500 channel-mixing network.

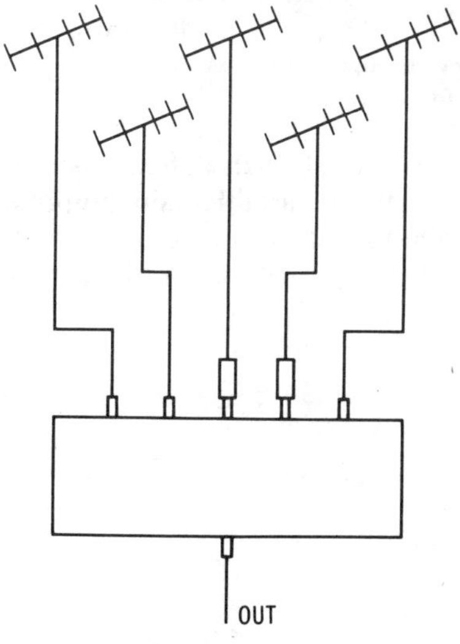

Fig. 8-19. Equalizing signal levels with a passive channel-mixing network.

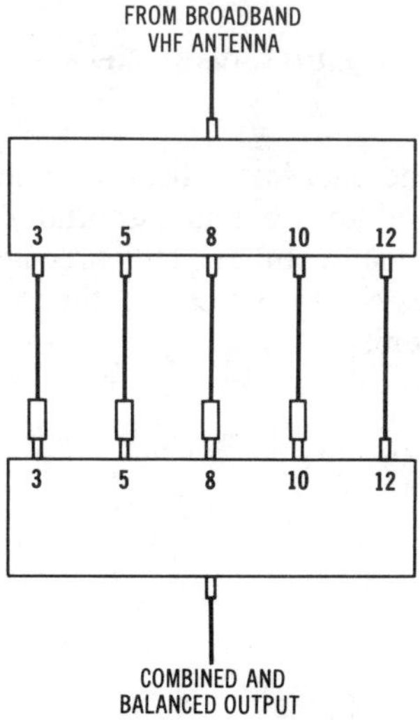

Fig. 8-20. Signal-level equalization of a broadband antenna by channel separation.

Channel-mixing networks can also be used to equalize the signal levels of channels provided by a broadband antenna. This technique is excellent for areas served by markets where all or most of the tv transmitting antennas are located at the same place, e.g., New York City, Philadelphia, Portland, Los Angeles, Denver, Des Moines. Two channel-mixing networks are required (Fig. 8-20): One is used backwards to separate the channels, the other recombines the equalized channels. Barrel attenuators between the two networks attenuate all but the weakest channel down to the level of the weakest channel.

Attenuation as a method of equalizing gain is excellent for strong-signal areas, but active techniques are needed for weak-signal areas.

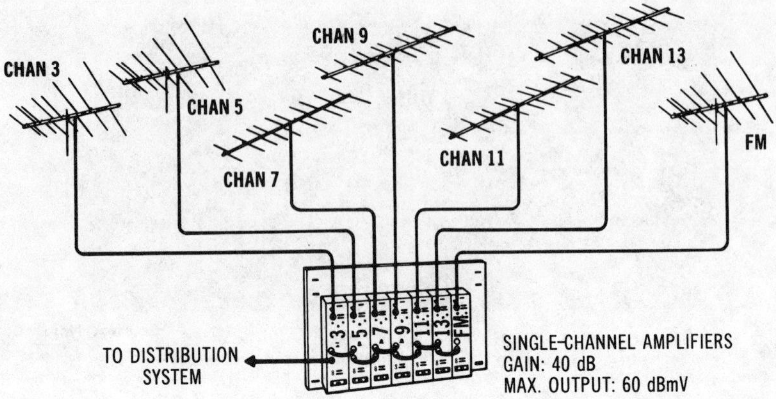

Courtesy TACO/Jerrold

Fig. 8-21. Catv/matv antenna system using separate antennas for each channel.

In the catv/matv head end shown in Fig. 8-21 each antenna is connected to what is essentially a combination variable-gain amplifier and yagi coupler. This arrangement allows each channel signal to be adjusted to precisely the level needed by the signal-distribution system.

CHAPTER 9

Designing the Antenna System

By this point in the book you have seen all of the background information needed to enable you to design your own antenna system. This chapter shows how to use all the previous information.

There are three basic steps to designing your antenna system. They are:

1. Find out what you are starting with. Make a signal survey to determine the signal levels at your reception site, then pick a suitable antenna.
2. Decide on what you want to end up with. Make a diagram showing the number of tv sets (and possibly fm tuners) you want to operate, and how you want to distribute the signal to them.
3. Figure out what equipment (if any) is needed to interface the antenna system of Step 1 with the signal-distribution system of Step 2.

SIGNAL SURVEY

In fringe areas you know beforehand that the signal levels are so low you need the biggest antenna practical, mounted on as high a tower as safety (or local ordinances) permits. For all other areas (metropolitan to suburban) making a signal survey before buying and erecting the permanent antenna is advisable.

To make a signal survey, mount a small- to medium-sized antenna equipped with a balun and coax on a 10-foot mast section, and measure the signal level on every channel you want to receive. Be certain to orient the antenna for maximum signal strength on each channel. Check the picture quality on a portable tv set while doing this, to see if ghosts are a problem on any channels. Use an antenna with good front-to-back ratio on all channels, and relatively flat gain characteristics. Excellent antennas for this purpose are the RCA 3BG09 for vhf, and the Gavin CR-5 for uhf.

There are two techniques for measuring the signal level: the field-strength meter technique and the attenuator technique. Both of these methods are described in the following sections.

Field-Strength Meter Technique

The field-strength meter is a frequency-selective voltmeter. This means it is capable of measuring the signal of each channel individually. "Signal-level meter" is actually the correct term for these devices, since in most matv applications they measure signal level rather than field strength. However, "field-strength meter" is the more popular term, and it will be used in this book.

A good field-strength meter for signal surveys can measure signal levels ranging from 50 microvolts (−26 dBmV) to 100,000 microvolts (+40 dBmV) in 75 ohms. It should have both microvolt and dBmV scales, and cover Channels 2–83. The Sadelco FS-719B shown in Fig. 9-1 is an excellent example of a high-quality, low-cost field-strength meter. In fact, Sadelco field-strength meters are sold by many tv antenna manufacturers, including Blonder-Tongue, Channel Master, Winegard, and Finco. The prices of these meters range from $300 to $550. Anyone installing antennas professionally or charged with the maintenance of an matv system should consider purchasing one of these.

To measure signal level, connect the coax from the test antenna to the input of the field-strength meter. Tune the field-strength meter to the first local channel and adjust the input attenuator for highest on-scale reading. Record the signal level in both microvolts and dBmV. Do this in turn for each channel you wish to receive.

Attenuator Technique

The attenuator technique is an economically practical alternative to the field-strength meter for the occasional installer or the home owner installing his own system. If a small (5- to 9-inch) tv set is available, a step attenuator and possibly one or two low-cost barrel attenuators are the only things that must be purchased. Traditional step attenuators like the Winegard SA-62 can produce 0–62 dB attenuation in 1-dB steps, hence are the most convenient to use in this procedure. The RMS Electronics CA-1122VA (Fig. 9-2) covers only 0-24 dB in 3-dB steps, but the unbeatable uhf performance of this miniature attenuator makes it the top choice when many uhf channels are involved. Its attenuation range can be extended with 20-dB barrel attenuators (the CA-1121M-20 shown in Fig. 6-24 for instance) if necessary.

The technique is based on the premise that most modern tv sets will become noticeably snowy when the signal level is about 180 microvolts (−15 dBmV) at its vhf 75-ohm input and 360 microvolts

Courtesy Sadelco, Inc.

Fig. 9-1. Sadelco FS-719B field-strength meter.

(−9 dBmV) at its uhf 75-ohm input. (If the set does not have 75-ohm inputs, connect baluns to the 300-ohm terminals.) The test setup is shown at the top of Fig. 9-3. Tune the tv to the first local channel. Adjust the step attenuator to add just enough attenuation to make the picture snowy. Record the amount of attenuation used, then read the equivalent signal level from the appropriate section of Table 9-1.

Fig. 9-2. RMS Electronics CA-1122VA step attenuator.

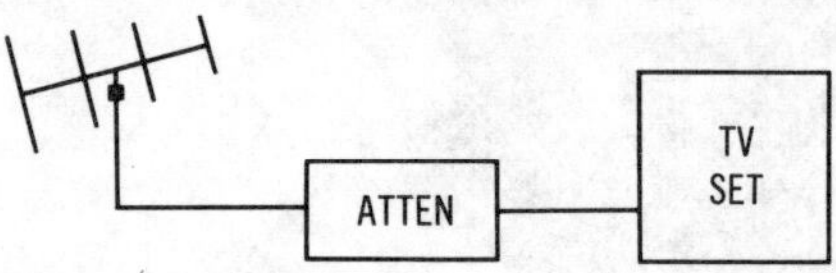

CHAN	ATTEN	SIGNAL LEVEL dBmV	μV	COMMENTS ON PICTURE QUALITY

Fig. 9-3. Measuring and recording signal level by the attenuator technique.

HEAD-END DESIGN

Head end is a catv term covering the "outside" components of the system: the antenna, transmission line, and mast-mounted preamp (if any). The basic idea in head-end design is to select the components needed to produce *at least* 500 microvolts in 75 ohms on each vhf channel, and 1000 microvolts on each uhf channel, at the point where the transmission line enters the building. (These same levels

Table 9-1. Attenuator Setting to Signal-Level Conversion

VHF Signal Level			UHF Signal Level	
dBmV	μV	Attenuation	dBmV	μV
−15	180	0	−9	360
−12	250	3	−6	500
−9	360	6	−3	700
−6	500	9	0	1,000
−3	700	12	3	1,400
0	1,000	15	6	2,000
3	1,400	18	9	2,800
5	1,800	20	11	3,600
8	2,500	23	14	5,000
11	3,600	26	17	7,000
14	5,000	29	20	10,000
17	7,000	32	23	14,000
20	10,000	35	26	20,000
23	14,000	38	29	28,000
25	18,000	40	31	36,000
28	25,000	43	34	50,000
31	36,000	46	37	70,000
34	50,000	49	40	100,000
37	70,000	52	43	140,000
40	100,000	55	46	200,000
43	140,000	58	49	280,000

are also minimum values for the signal levels delivered to each tv set.) In practice, figures of 1000 microvolts vhf and 2000 microvolts uhf are used to allow for variations in propagation, or tv receivers with poor sensitivity.

Antenna Selection

If the signal survey with the RCA 3BG09 indicates signal levels of 1000 microvolts (0 dBmV) or higher on all vhf channels, any of the vhf antennas featured in this book will do as far as antenna gain is concerned. In this case you can do with the smallest antenna having the pattern characteristics you need (for ghost and interference rejection). If ghosts are a problem, you might need a large antenna because of its better front-to-back ratio and narrower beamwidth. Or, in very strong signal areas (over 10,000 microvolts per channel from a medium-sized antenna) you might find that the 2–3 dB higher gain of a large antenna is just enough to avoid using a distribution amplifier in the signal-distribution system.

If the signal survey indicates signal levels of 500 to 1000 microvolts (—6 to 0 dBmV), a medium-sized vhf antenna like the Winegard CH-4052 or a high-gain uhf antenna like a 4-bay bowtie will do. Levels of 200 to 500 microvolts (—14 dBmV to —6 dBmV) measured on all (or most) channels during the signal survey require a very large antenna for vhf, and an array of high-gain antennas (side-by-side 4-bays, for instance) for uhf, if a Grade A picture is desired at all times. Lower levels (under 200 microvolts, or −14 dBmV) require the largest-possible antenna and the signal-optimization techniques discussed in Chapter 8.

As an example of what kind of antenna fits what signal levels, Table 9-2 matches appropriate commercial antenna models to various signal levels. Note that the selections in this table are intended *strictly as an example;* other antennas covered in this book, and many not covered, can be used instead. The reason this is so is because the selections in Table 9-2 were made on the basis of the *average* gain.

Table 9-2. Antennas Suitable for Various Signal Levels

Band(s)	5000 μV (+14 dBmV)	2000 μV (+6 dBmV)	1000 μV (0 dBmV)
VHF	Winegard CH-4210	Winegard CH-4210 (except in Chan 2 areas)	RCA 3BG09 Antennacraft Mark-8 RMS DJR-6 A.C.A. AC-511
UHF	JFD UHF-600	Gavin CR-5 RCA 7B140 Jerrold PAU-700	Ch. Master 4247
VHF-UHF		RCA 4BG15 Antennacraft CDX-650 A.C.A. ES-270	RCA 4BG20 Ch. Master 3676A Jerrold VU-932S
Band(s)	**500 μV (−6 dBmV)**	**200 μV (−14 dBmV)**	**100 μV (−20 dBmV)**
VHF	Winegard CH-4052 Jerrold VIP-303 Winegard CH-4053 Jerrold VIP-304	Winegard CH-5200 Jerrold VIP-307	Same as previous column plus low-noise mast-mounted preamplifier
UHF	JFD UHF-202 Lance KW4S A.C.A. AC-320 Winegard KU-420 Antennacraft G-1483	Same as previous column in stack of two	Same as previous column plus low-noise mast-mounted preamplifier
VHF-UHF	Jerrold VU-933S Jerrold VU-934S	Separate UHF & VHF antennas as above	Separate UHF & VHF antennas as above

However, an antenna whose *average* gain is less than that of another antenna may have higher gain on the few channels you want to receive. In fact, the gain variation on an individual-channel basis is far greater than the average gain variation between antennas. For this reason it is very important to determine the antenna performance *on the weakest channel in your area,* either via the gain data in this book, or from manufacturers' data. Then select the antenna with the highest gain on the *weakest channel* in your area. Another factor is transmission-line length. If many channels of about the same strength are to be received and the line run is very long, an antenna with a rising gain characteristic helps in equalizing the signal levels at the output end of the transmission line. For example, the Jerrold ZIP-12V and Winegard CH-4210 are suitable in this application for vhf reception, as are the Lance KW4S and RCA 7B140 for uhf reception.

The physical construction of the antenna should also be considered. Both vhf antennas and the vhf sections of vhf-uhf antennas are very susceptible to damage during ice and wind storms, particularly the long elements snapping off near the boom. When choosing an antenna for a severe climate, disregard manufacturers' claims about boom and insulator strength; failures rarely occur nowadays at these points. Instead, check the element attachment to the boom insulator. Look for reinforcements such as wood or plastic plugs *inside* the element, and metal sleeves outside that extend over the element for several inches out from the boom insulator. Another point to check is the mast clamp. Be sure it is a multipart affair that completely surrounds the boom; simple U-bolts will crush (hence weaken) the boom. Also make sure it is capable of a tight and rigid fit on the mast you are using.

Last, but certainly not least, is pattern characteristics. Obtain a map of your area, and on it mark your location and those of the stations you wish to receive. This will show the relative bearing between stations. If the station bearings are close enough together, you can avoid using a rotator by picking an antenna whose −3-dB beamwidth is greater than the maximum difference in bearing. For vhf-H channels, this means an antenna multimoded by extension stubs, such as the RCA 3BG09, 4BG15, or Winegard CH-4052; for uhf channels this means a plane or corner-reflector antenna (see Table 3-4). However, if you have a high-band adjacent-channel interference problem, and the interfering station is in almost the same direction as the desired one, narrow beamwidth is needed. For the vhf high band an antenna multimoded by parasitics or vee'ing is preferable; then you can adjust the antenna orientation to drop the interfering station into the null adjacent to the front lobe. Antennas such as the Antennacraft Mark-8 and CDX-650, RMS Electronics DJR-6, or Jerrold VIP-303 are suitable for the vhf-H application. Pattern is rarely a consideration for vhf low-band reception, since all types of small- to medium-sized broadband vhf antennas have wide beamwidths on the low band. The only exception is when a sharp side null is needed to knock out interference coming from that direction; then antennas like the RMS Electronics DJR-6 and Jerrold VIP-303 or VU-932S are indicated.

Transmission-Line Selection

The advantages and special applications of the various types of transmission lines were discussed in Chapter 5. The basic facts can be summarized as follows:

1. For *any* antenna system using a rotated antenna and/or serving several tv sets, use foam-dielectric coaxial cable. RG59-sized coax is sufficient for most vhf-only installations, and for strong-signal uhf areas. RG6-sized coax should be used for uhf and vhf-uhf installations in weak-signal areas, and for very long line runs (over 300 feet) at any frequency. RG11-sized coax is generally reserved for special applications, an example of which is given later in this chapter.
2. Twinlead can be used, primarily for convenience, when the antenna is fixed (nonrotating) *and* only one tv receiver is involved. In extremely dry climates the twinlead can be tubular foam (Belden 8275) or oval foam-encased (Belden 8285). Shielded twinlead is recommended for all other climates: the small size (Belden 9090) for vhf-only areas, the large size (Belden 8290) for uhf and weak-signal vhf areas.

The factors involved in selecting line size (i.e., RG59 vs. RG6, Belden 9090 vs. 8290) are the signal levels at the antenna, the length of the line run from the location of the permanent antenna to the tv receiver or distribution system, and the frequency of the highest channel to be received. One channel nearly always determines the minimum-usable line size. It may be the weakest channel, or the highest in frequency, or both. "Worst-case" calculation eliminates the need to decide difficult cases by figuring cable attenuation at the highest channel carried, and using the lowest-level channel for determining the resultant output level. Keep in mind, though, that in some cases this can cause you to put more money and effort into the antenna system than is really necessary.

The basic technique is to look up the loss per hundred feet for the desired cable at the highest frequency carried. Use the 100-MHz attenuation figures in Table 5-2 for systems carrying only low-band vhf channels, or the 200-MHz figures for a vhf-only system with high-band stations. For uhf or vhf-uhf systems, use the 500-MHz attenuation figures if most of the uhf channels are at the lower portion of the band, or the 900-MHz figures if weak channels are at the upper portion of the band or translator channels are involved. Multiply the loss per hundred feet figure by the proposed length of the transmission line, then divide the result by 100. The final answer is the dB loss of your transmission line. Subtract the dB loss thus calculated from the signal level (in dBmV) of the weakest channel. For instance, suppose Channels 2, 5, 7, and 13 produce over 10,000 microvolts each at the antenna terminals, and Channel 20 produces only 5000 microvolts. Obviously, Channel 20 is the one to use, since it is both the highest in frequency and the weakest. Channel 20 is close to the bottom of the uhf band; so the 500-MHz attenuation figures are the ones to use. Seventy-five feet of RG59-sized foam coax has $(6.2 \times 75) \div 100 = 4.6$ dB loss. Subtracting this from +14 dBmV (5000 microvolts) yields +9.4 dBmV output from the transmission line. This is above the +6 dBmV level desired for uhf, so RG59-sized line is okay in this system. If this uhf station was weaker, say +9 dBmV (2800 microvolts) at the antenna, the loss of 75 feet of this cable would drop the output level to +4.4 dBmV, below that desired for directly driving a tv set. You would then have to use either a larger-sized line (RG11 foam) or a low-noise indoor preamplifier (such as the Winegard UA-4050). If the uhf station was very weak, −6 dBmV (500 microvolts) for instance, there is no way line selection could help, since the antenna output level is already below the desired level. In this case low-noise amplification *at the antenna* is necessary. Since the preamplifiers recommended in Chapter 6 provide 15–18 dB gain, any line can be used after the preamplifier unless the line run is very long (over 200 feet). RG59-sized foam line (Belden 9275 and equivalents) is the usual choice because of its low cost.

The choice of the transmission line increases in importance as the length of the transmission line increases, and as the channel frequencies involved increase. For very short line runs (30–40 feet), the difference in attenuation between RG59 foam and even RG11 foam is insignificant at vhf frequencies and fairly small even at uhf frequencies. But for very long runs (several hundred feet) the difference is substantial. For instance, if the line run in the first example was 400 feet (antenna mounted on nearby hilltop), the 25-dB attenuation of RG59-sized foam would reduce the output level to −11 dBmV (too low to yield a noise-free picture under any circumstances), while the 15-dB attenuation of RG11 foam coax would lower it to only −1 dBmV, a level which will yield a very good uhf picture on *most* tv sets.

Preamplification

If the expected signal levels at the output of the transmission line drop below 500 microvolts (−6 dBmV) at vhf frequencies or 1000 microvolts (0 dBmV) at uhf frequencies, a mast-mounted preamplifier should be installed. This will provide the most snow-free picture possible under the circumstances. The longer the line run, the more important a *mast-mounted* preamplifier is. Similarly, the higher the channel frequencies, the more important a mast-mounted preamplifier is.

At uhf frequencies a low-noise preamplifier is worth its weight in gold, regardless of the length of the line run. This is because the very low noise figures of the preamplifiers recommended in Chapter 6 allow a 300-microvolt signal to produce the same Grade A picture quality as a 1000-microvolt signal applied directly to the uhf tuner. If the line run is long, a mast-mounted preamplifier is even more valuable, since the effect of transmission-line loss on signal-to-noise ratio is substantially eliminated.

At vhf frequencies the improvement is significant but not as dramatic, because the difference in noise figure between a good preamplifier and tv-set tuner is not as large as it is at uhf frequencies. The best of the vhf preamplifiers recommended in Chapter 6 will allow a 300-microvolt signal to produce the same Grade A picture as a 500-microvolt signal applied directly to the average vhf tuner. Three-hundred microvolts at the *antenna* terminals is therefore the signal level to try for in fringe and far-fringe areas when mast-mounted preamplifiers are used.

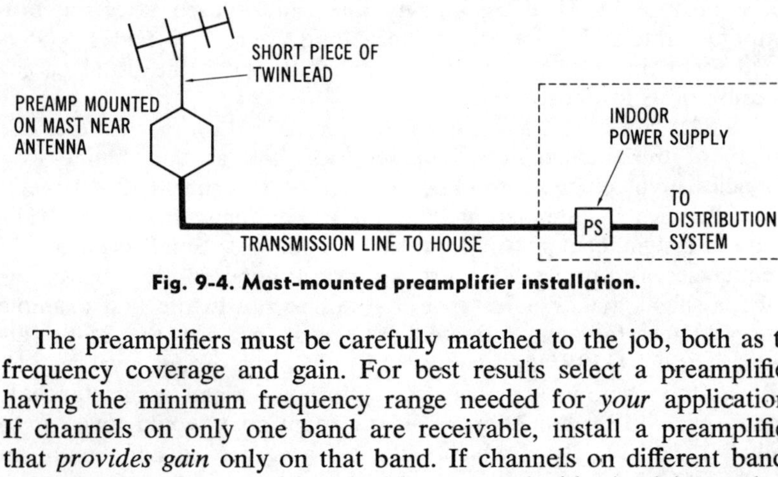

Fig. 9-4. Mast-mounted preamplifier installation.

The preamplifiers must be carefully matched to the job, both as to frequency coverage and gain. For best results select a preamplifier having the minimum frequency range needed for *your* application. If channels on only one band are receivable, install a preamplifier that *provides gain* only on that band. If channels on different bands (vhf-H and vhf-L, or vhf and uhf) are receivable (and in need of amplification) use a preamplifier with the matching combination of bands. The variety of preamplifiers listed in Chapter 6 makes this possible. Similarly, the gain of the preamplifier should match the situation; use a medium-gain (around 15 dB) preamplifier for fringe areas, and high-gain preamplifiers (around 25 dB) for far-fringe areas or installations with unusually long line runs. Do *not* use an all-channel preamplifier, particularly a high-gain model, if the signals on one band or the other are strong. Use a preamplifier that provides gain *only* for the weak band, but passes the strong-band signals to the transmission line without amplification.

Mount the preamplifier on the mast close to the antenna, and connect the preamplifier input and antenna output terminals with a short piece of twinlead, as in Fig. 9-4. (Any kind will do for this application.) The preamplifier is powered through the same coaxial cable that carries signal to the set. The indoor power supply has a signal output connector to which the tv set or distribution system is connected.

DISTRIBUTION-SYSTEM DESIGN

Distribution-system layout is often a difficult task for the beginner because there are a tremendous number of possible layouts. The number of tv receivers served, their location, the arrangement of rooms in the structure being wired, the signal levels from the head end, the frequencies involved, and just plain philosophical differences make each signal-distribution system unique. However, once certain concepts and techniques are understood, distribution-system layout for houses and small matv systems is a snap! To this end you will be taught the basic layouts and calculation techniques, and the major variations. You will also follow the step-by-step design of a typical

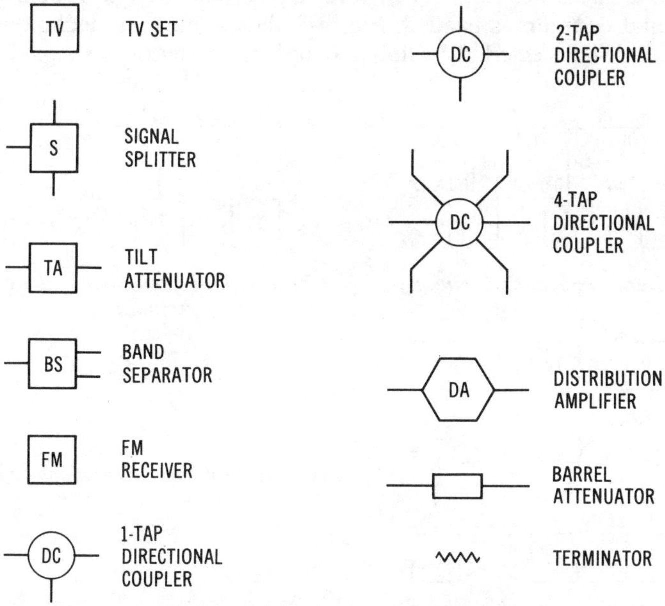

Fig. 9-5. Legend for symbols used in systems diagrams of Chapter 9.

small matv system. In all these cases the schematic diagrams of the layouts and systems will use the symbols shown in Fig. 9-5.

System Layout

The first thing to do is decide on how many outlets are needed. If only a single set is involved, the transmission line is simply interfaced with the tv set as described later in this section. If only two sets are involved, and top performance is not the most important consideration, the "quick and dirty" technique shown in Fig. 9-6 can be used instead. Only +3 dBmV (1400 microvolts) is needed at the input to the two-way signal splitter to produce the desired 0 dBmV (1000 microvolts) at each tv set for vhf channels. For uhf

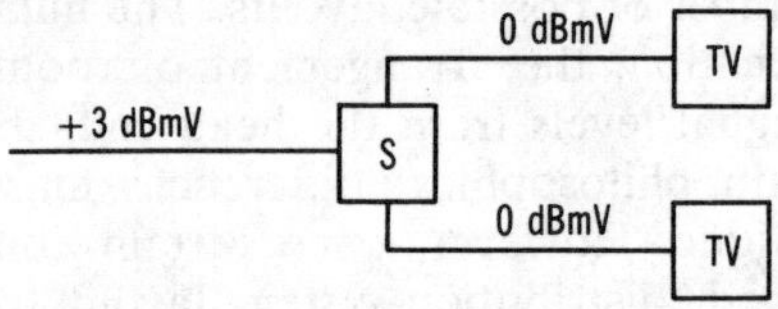

Fig. 9-6. Feeding two sets with a signal splitter.

signals, +9 dBmV (2800 microvolts) is required at the input to produce +6 dBmV (2000 microvolts) at each tv set.

When more than two sets are involved, and/or a first-class distribution system is desired, a high-level signal-distribution system using directional couplers is needed. Fig. 9-7 shows the basic technique for using series-connected directional couplers to distribute signal to a

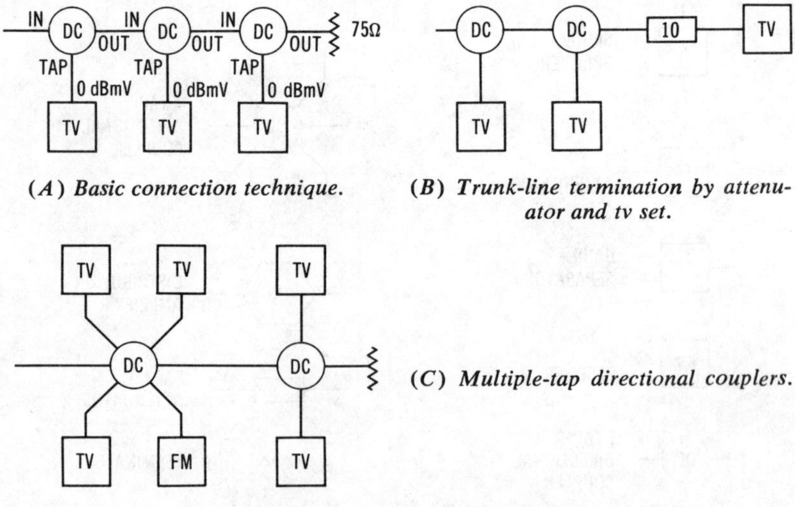

(*A*) *Basic connection technique.*

(*B*) *Trunk-line termination by attenuator and tv set.*

(*C*) *Multiple-tap directional couplers.*

Fig. 9-7. High-level distribution system.

large number of tv receivers. For clarity, only a few directional couplers are shown in these diagrams, although the technique is applicable to dozens of them.

The basic technique is shown in Fig. 9-7A. One directional coupler is used for each tv set being supplied. A length of coax connects the output port of one directional coupler to the input port of the next one down the line. The length of these cable sections may be only 10 feet or so, or hundreds of feet, depending on the distance between the sets. The tap port of each directional coupler is connected with coax to an outlet plate serving the tv set, or directly to the tv set itself.

A 75-ohm terminator is connected to the output port of the last directional coupler in the string. This allows a convenient place to make level checks on the system without interrupting anyone's service. If desired, the last directional coupler can be replaced by a 10-dB barrel attenuator (which is usually cheaper than a directional coupler), as in Fig. 9-7B.

When several tv sets are located fairly close together, it is more economical (and less work) to use a multiple-tap directional coupler (Fig. 9-7C). For instance, in motels and many residences, four rooms have a common corner juncture. A four-tap directional coupler in the ceiling at this point handily serves four tv sets, even though they are located in four different rooms.

The signal-distribution system can also distribute fm signals if the antenna feeding the system is capable of picking them up. Normally, no system modifications are required; simply connect the fm tuner to the tap outlet instead of a tv set. However, if certain local fm stations are so strong as to cause fm interference on some of the tv sets, it will be necessary to install an fm bandstop filter at the input to each tv set so affected.

The series-connection technique shown in Fig. 9-7 employs one trunk line, which is directly connected to the head end (or an amplifier). This is more than adequate for private residences, even two-story private homes. However, for motels and apartment houses it is highly impractical to connect everything in series, since a failure in one of the early directional couplers will cause a signal loss to everything thereafter. For larger systems the series-parallel technique is best. The signal from the head end or distribution amplifier is divided into two or more branches by a signal splitter, and each output of the signal splitter drives a string of directional couplers (Fig. 9-8). This technique is especially good for apartment houses (where a separate branch is used for each floor) and motels (where a branch is used for each wing). Eight-output splitters are available for structures having more than four floors or wings, and additional splitters can be attached to the outputs of an eight-way splitter if needed for a larger number of branches. The series-parallel arrangement provides additional benefits. If all branches are identical, only one set of tap-value calculations (discussed in the next section) need be made. On the other hand, if one or more of the branches are dissimilar (penthouse apartments on a hotel, for example), or one branch serves an area far from the others, amplification can be individually applied where needed. A good example of this is shown in the four-branch system in Fig. 9-8B; the lowermost branch serves an outbuilding (annex, guard shack, servants' quarters, etc.) remote from the main building. A distribution amplifier and tilt attenuator are used to compensate for signal loss in the long cable run from the main house.

The best location for a distribution amplifier used as a booster for long cable runs depends on several factors. Placing the distribution amplifier before the cable run yields the best possible signal-to-noise ratio, but cross modulation may result if the signal level at the splitter output is high and the amplifier gain is high (low maximum input level). Placing the distribution amplifier at the end of the cable (as in Fig. 9-8) virtually eliminates the possibility of cross modulation, but may degrade the signal-to-noise ratio of the signal delivered to that branch. The deciding factors are the signal levels out of the splitter, the maximum input level of the distribution amplifier, the gain of the distribution amplifier, and its noise figure. In Fig. 9-8 the reasoning was that since the signal level at the splitter output must be fairly high (+16 to +24 dBmV) to drive the unamplified branches, even 12 dB of cable attenuation (over 250 feet at vhf frequencies) would leave the signal level high enough for a noise-free picture, particularly if a low-noise (3-dB noise figure) distribution amplifier was used.

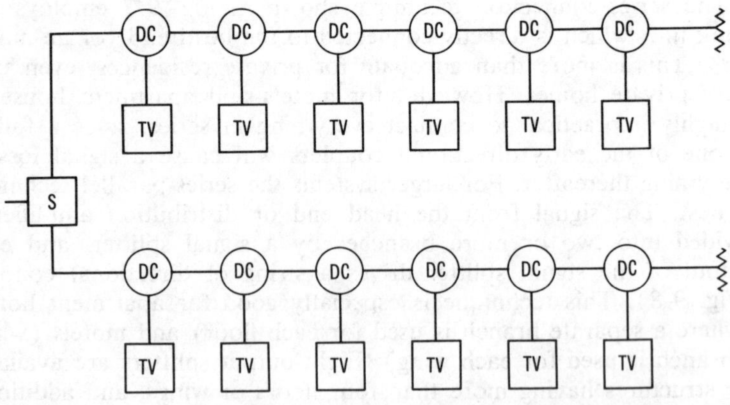

(A) Two-branch system.

Fig. 9-8. Multibranch

Calculating Tap Values

After sketching the system, the tedious part of the job begins. Mark the individual lengths of the coax used to connect the directional

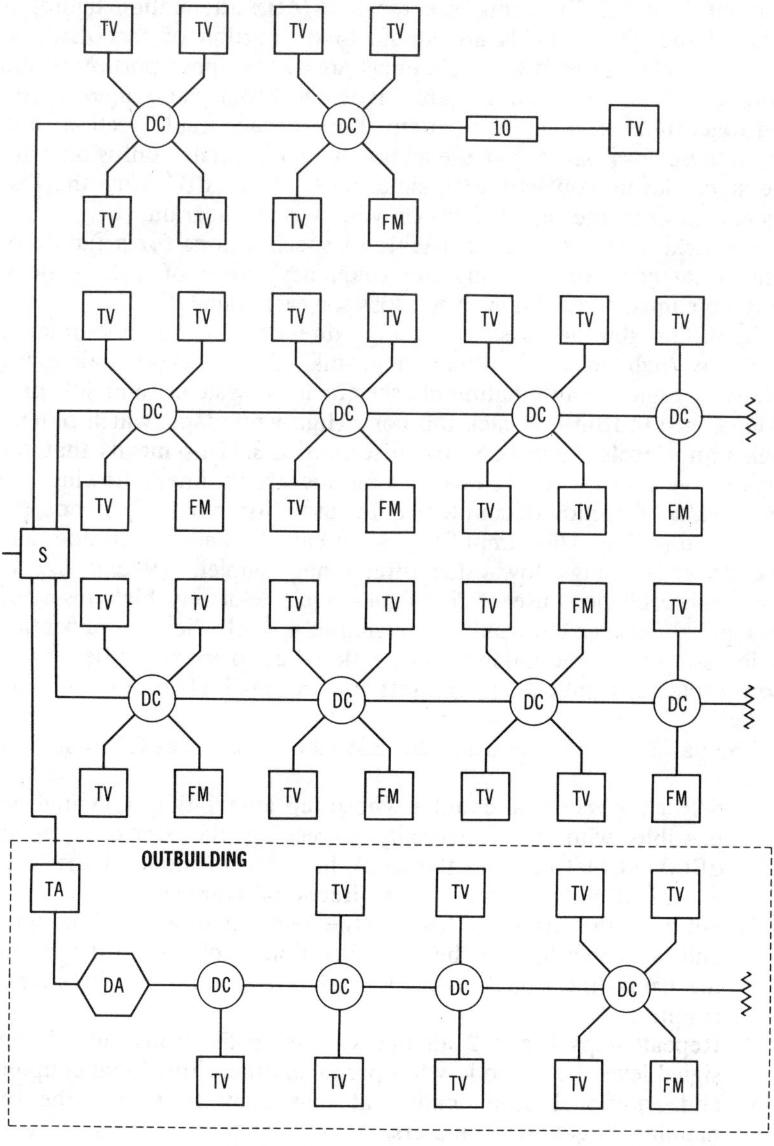

(B) Four-branch system.

distribution systems for matv.

couplers. Then look up the loss per hundred feet for the cable being used at the highest frequency carried. Use the 100-MHz attenuation figures in Table 5-2 for systems carrying only low-band vhf channels, or the 200-MHz figures for a vhf-only system with high-band stations. For uhf or vhf-uhf systems, use the 500-MHz attenuation figures if most of the uhf channels are at the lower portion of the band, or the 900-MHz figures if weak channels are at the upper portion of the band or translator channels are involved. From these figures you can determine the maximum attenuation of each cable section, and mark these loss figures on the diagram. To prevent confusion with the signal levels you will soon determine, write "dB" after the loss figures, and *box* the signal levels on your system diagram.

The next step is to secure a table of specifications for a family of directional couplers covering the frequency range of your system. The table must show the insertion loss for each value.

Signal-distribution systems using directional couplers require relatively high levels all along the trunk line to accommodate the relatively high tap attenuation chosen for most systems, and still provide *at least* 0 dBmV at each tap port. High-value taps, which require high input levels, have very low insertion loss. This means that the signal levels drop very slowly along the trunk line, allowing the same value directional coupler to be used for most of, if not the entire, trunk line. This simplifies system calculations, purchasing, and inventory. Although low-value directional couplers (9 and 6 dB) make more efficient use of the signal, their relatively high insertion loss quickly erodes the trunk-line signal to a level where amplification is necessary if a large number of directional couplers are to be served. Low-value taps have their greatest use in small (two- to four-set) systems.

The basic technique in using directional couplers is as follows:

1. Select a directional coupler whose tap attenuation is as high as possible without numerically exceeding the signal level in dBmV at that point in the trunk line. For example, if the level is +27 dBmV, select a 24-dB directional coupler.
2. Subtract the insertion loss of the selected directional coupler and the cable loss to the next directional coupler to determine the trunk-line signal level at the input to the next directional coupler.
3. Repeat Steps 1 and 2 all the way down the trunk line. If the signal level drops too low to operate another directional coupler, add additional amplification at that point to supply the remaining directional couplers.
4. Connect a 75-ohm terminator to the OUT port of the last directional coupler in the string.

As an example of how to lay out a system and calculate the component values, let's design a system for a two-story private residence with an outbuilding (Fig. 9-9). This arrangement fits situations from "mother-daughter" homes with a finished garage, to the "richman's" residence, a two-story town house with detached servants' quarters. (One can dream, can't one?) To keep from excessively complicating the example, this will be a vhf-only system.

The signal survey with the RCA 3BG09 shows the levels listed in Table 9-3. A quick look shows Channel 5 to be the weakest channel, and Channel 7 the strongest. Although Channel 13 is 1 dB stronger

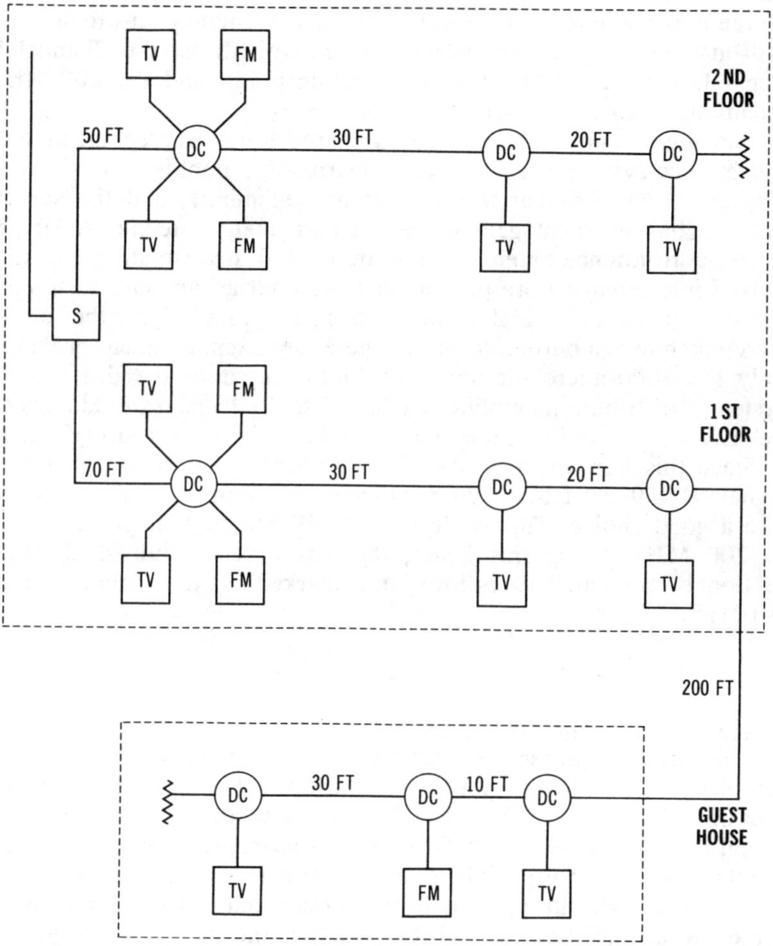

Fig. 9-9. System layout for calculation example.

Table 9-3. Results of Signal Survey for Design Example

Channel	Microvolts	dBmV
2	20,000	+26
4	28,000	+29
5	16,000	+24
7	36,000	+31
9	20,000	+26
11	22,000	+27
13	18,000	+25

than Channel 5, however, it is so very much higher in frequency (211 MHz vs. 77 MHz) that it will suffer much greater attenuation by the long cable runs. To cover all bases and thereby ensure *at least* 0 dBmV on every channel at each tv set, we will use the Channel 5 signal level (+24 dBmV) as the starting point, and the 200-MHz attenuation figure for calculating cable losses.

The high signal levels measured in the signal survey mean two things: The system will *not* need a distribution amplifier to drive the directional couplers (at least not at its beginning), and the survey antenna has sufficient gain to be the permanent antenna. A larger, higher-performance antenna would be needed in a situation like this only if interference were present that required an antenna with very narrow beamwidth or high front-to-back ratio. Similarly, if the survey were taken in a suburban location where the antenna system provided only the 1000-microvolt minimum at the input to the distribution system, a distribution amplifier having 15 to 30 dB gain would have to be installed here to bring the signal into the +20- to +40-dBmV range.

Since this is a vhf-only system, and there seems to be plenty of signal, RG59-sized foam coax (Belden 9275 or equivalent) looks like a good choice. This cable has 3.8 dB attenuation per 100 feet at 200 MHz. Using this figure, the loss of each individual cable section is calculated as follows, and marked on the diagram (Fig. 9-10):

$$\text{Loss (dB)} = \frac{3.8\,L}{100}$$

where L is the section length in feet.

The two-way splitter has 3 dB loss to each branch, so the minimum signal level at its output ports is +21 dBmV. The 1.9-dB loss of 50 feet of 9275 leaves +19.1 dBmV at the input to the four-tap directional coupler on the second floor. RMS Electronics taps were chosen for this system because of their superb vhf performance and low cost. If this were a vhf-uhf system, Jerrold directional couplers (see Chapter 6) would be used here. Table 9-4 lists the pertinent design criteria for the single- and four-tap families of RMS Electronics direc-

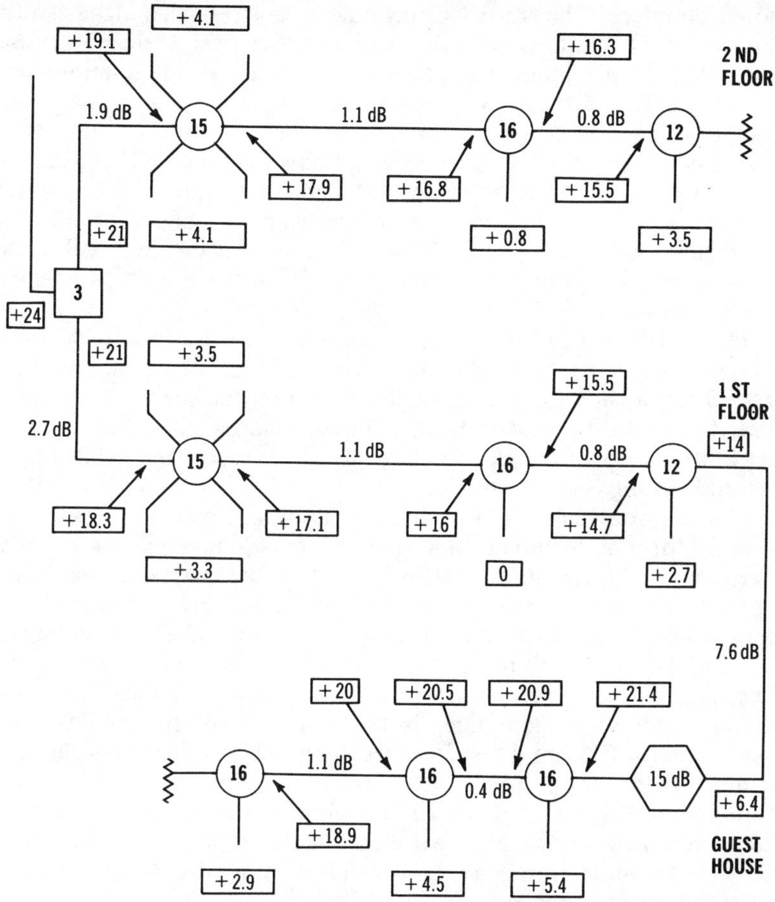

Fig. 9-10. Example system with losses and levels marked down.

Table 9-4. Design Specifications for RMS Electronics Directional Couplers

CA-1090M Single-Tap Series		CA-2014T Four-Tap Series	
Tap Attenuation (dB)	Insertion Loss (dB)	Tap Attenuation (dB)	Insertion Loss (dB)
6	2.0	—	—
9	1.2	10	3.5
12	0.7	—	—
16	0.5	15	1.2
20	0.4	20	0.7
24	0.4	25	0.5
30	0.4	30	0.5

tional couplers. The 15-dB directional coupler is the highest-value four-tap directional coupler that can produce over 0 dBmV output for +19.1 dBmV input. Subtracting the 15-dB tap attenuation from +19.1 yields +4.1 dBmV at each tap. Subtracting the directional coupler's insertion loss of 1.2 dB leaves +17.9 dBmV at its output port. Thirty feet of cable cuts the signal down another 1.1 dB, leaving +16.8 dBmV at the input to the first single-tap directional coupler. The CA-1090-16 is the obvious choice here. The signal at the next tap is too low for a 16, so a 12-dB unit is used at the end of the line. A 75-ohm terminator is connected to the OUTPUT port of the last directional coupler.

The first-floor trunk line is identical, but the extra loss in the extra 20 feet of cable run from the attic means that all the signal levels are 0.8 dB lower than the corresponding levels on the second-floor trunk line. The real difference though, is in what happens at the output of the last directional coupler. Instead of a terminator, the cable run to the outbuilding is connected here.

The 200-foot cable run to the outbuilding drops the signal level to +6.4 dBmV. Although a 6-dB unit could be used for the first directional coupler, its 2-dB insertion loss would leave insufficient signal to operate the remaining directional couplers. Amplification is one answer. A low-cost distribution amplifier like the Winegard DA-205 boosts the signal to the first directional coupler by 15 dB. The signal-to-noise ratio is not significantly degraded since the +6.4-dBmV input to the amplifier is fairly high, and the DA-205 is a low-noise (3-dB) amplifier. The resulting +21.4-dBmV output level is more than enough to drive the remaining portion of the system. Although the signal level is high enough to permit using 20-dB directional couplers for the first two directional couplers after the amplifier, 16-dB units were chosen instead to minimize the number of different values in the system.

To "put the icing on the cake" a tilt attenuator can be installed ahead of the amplifier to attenuate excessively high low-band signals (Channel 4 in particular). The difference in cable loss between the frequency of Channel 4 and 200 MHz is almost 1.6 dB per 100 feet, which is about 5 dB for the 320 feet of cable between the splitter and amplifier. Since Channel 4 started out 4 dB higher than Channel 13, it is 9 dB stronger than 13 at the input to the amplifier. Even Channel 5, which started out 1 dB lower than 13, is now 4 dB higher. These levels are well below the maximum per-channel input level of the DA-205 (+31 dBmV), so a tilt attenuator is not really necessary. However, for the purpose of the example, a 5-dB unit like the RMS Electronics CA-2200-5 would be selected to lessen the difference between high-band and low-band signals that are provided to the outbuilding.

There is an alternative in this example to using an amplifier for the outbuilding, one which also eliminates deciding whether or not to use the tilt attenuator. Studying the signal levels in Fig. 9-10 reveals that a good deal of signal is lost in the long cable runs (i.e., from the attic splitter to the first floor, and from the main building to the outbuilding. If the RG59-sized cable is replaced with one having much lower attenuation, perhaps the signal level arriving at the

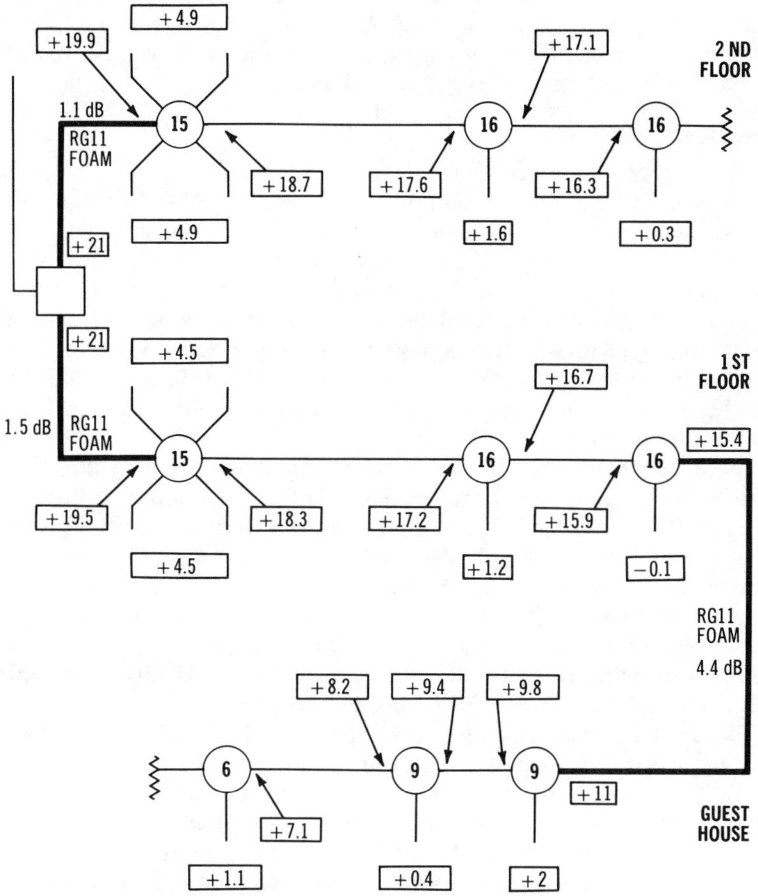

Fig. 9-11. Alternative system without amplifier.

outbuilding will be high enough to operate directional couplers without amplification. Fig. 9-11 shows the system levels recalculated with RG11-sized foam coax used for the long runs. The result is that the signal level at the input to the outbuilding trunk line is +11 dBmV vs. +6.4 dBmV for the all-RG59-sized system, allowing low-tap-attenuation directional couplers to supply the sets in the outbuilding. The lower attenuation of the RG11-sized cable also means that the difference in signal level between the high-band and low-band channels at the outbuilding is far less than it was in the all RG59-sized cable system.

Only the long cable runs were replaced with RG11-sized coax for two reasons: it wasn't necessary (in this case) to use it everywhere, and RG59-sized cable is much easier to work with. However, if you *care* to make the entire system of RG11, or if it is *necessary* to make the entire system of low-loss coax, go right ahead. You have paid for a 500-foot spool, so you might as well use it!

Receiver Interface

All tv receivers made in the U.S. have 300-ohm vhf and uhf input terminals. For a simple antenna system using 300-ohm line and carrying channels on only one band (vhf *or* uhf) the transmission line is simply connected to the appropriate input terminals. If both vhf and uhf channels are supplied by the antenna system, a 300-ohm band separator must be connected between the transmission line and the tv set input terminals. Be sure the band-separator output leads are connected to the proper set of tv terminals; attach the short stiff leads to the uhf input terminals, attach the long flexible leads to the vhf input terminals.

When coaxial cable is used, a band separator with 75-ohm input and 300-ohm output impedance is needed if the antenna system supplies both vhf and uhf channels. If channels in only one band are received (or desired), a balun is used instead of the band separator to match the coax to the appropriate tv input terminals. A balun is *always* required for the uhf input, and *usually* required for the vhf input. However, many new tv sets have an auxiliary 75-ohm vhf input connector. In this case coax carrying only vhf channels can be connected directly to the 75-ohm connector.

Baluns and band separators suitable for each of the aforementioned situations were discussed in detail in Chapter 6.

When a signal-distribution system with directional couplers is used, each tv set must be connected to the outlet serving it with a short length of flexible cable. Four to six feet are the most popular lengths. Many tv manufacturers sell prefabricated cables in different colors for this purpose. If you make your own, use an RG59-sized foam cable with *No. 22* center conductor and *copper-braid* shield. Belden 8221 is excellent for this purpose. The 80-ohm impedance resulting from a No. 22 center conductor with foam dielectric is unimportant in the short lengths involved. What *is* important is its high flexibility and ruggedness, because these cables receive a lot of abuse when the tv set is moved or the room is cleaned.

MATCHING THE DISTRIBUTION SYSTEM TO THE HEAD END

Once you have calculated the minimum input level required by your signal-distribution system, and have estimated the signal levels that will be available from the head end (antenna system), the only design job left is reconciling the difference, if necessary. (NOTE: If you erect the antenna system first, you can use measured signal levels rather than estimated values.)

The reconciling procedures may involve amplification, attenuation, or nothing more than connecting the head end and signal-distribution system together. The situation is generally one of the following:

1. If the signal produced by the head end is lower than that needed by the distribution system, a distribution amplifier is needed to boost the head-end output to equal or surpass the input level required by the signal-distribution system. This is generally the case whenever a high-level signal-distribution system (directional couplers) is used anywhere other than very close to the transmitter.

   Most of the time a distribution amplifier covering the appropriate frequency range is simply connected between the head end and distribution system. (Suitable distribution amplifiers were shown in Chapter 6.) However, if the channels in one band are very much stronger than those in another, one of the "tricks" shown in Fig. 9-12 can be used. The setup in Fig. 9-12A uses H/L band separators to bypass the vhf-L signals around the distribution amplifier, while the vhf-H signals are amplified. This technique is useful when the levels are too far apart for a tilt attenuator to fix, or for situations where the vhf-H channels are the strong ones. (In this latter case the distribution amplifier would naturally be inserted in the vhf-L path.) When only uhf signals require additional amplification (which is often the

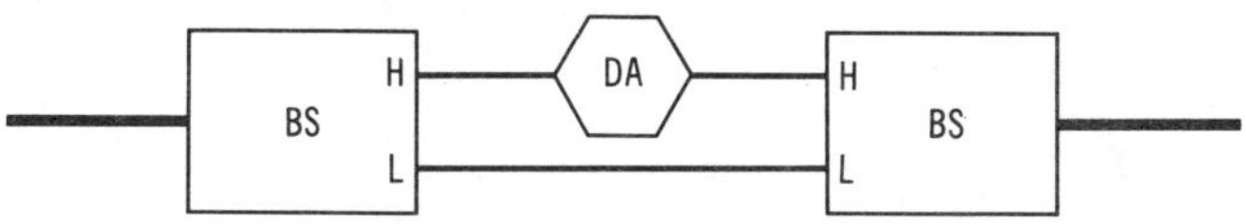

(*A*) *With vhf-H amplification.*

(*B*) *With uhf amplification.*

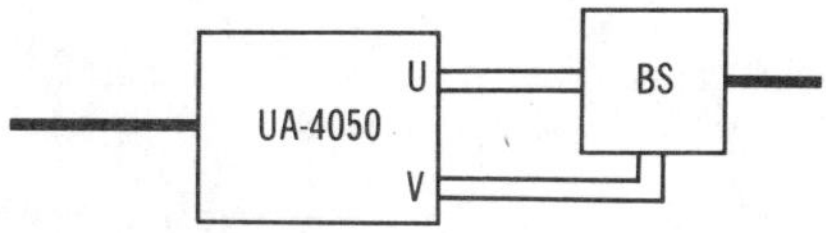

Fig. 9-12. Bypassing schemes.

case), the setup of Fig. 9-12B is used. The low-noise Winegard UA-4050 has a 75-ohm band separator at its input, so to use this device as a distribution amplifier with bypassed vhf frequencies, simply combine the uhf and vhf outputs with a 75/300-ohm band separator.

2. If the signal levels at the head end exceed those required by the signal-distribution system by a moderate amount (0–30 dB), you can simply connect the signal-distribution system directly to the head end without any level adjustment.

   Signal levels high enough to drive a high-level distribution system for a home or small matv system are often available from medium-sized antennas less than 10 miles or so from the transmitter.

3. If the signal levels at the head end are over 30 dB higher than those required by the signal-distribution system, attenuation is advisable.

   A situation like this may occur when a large antenna must be used within a few miles of the transmitter (because of multipath or some other problem), and only one tv set is served. The signal level delivered to the set may then be high enough to cause overload. Fortunately, the cure is simple; insert a barrel attenuator between the antenna output and tv set. Screw the attenuator directly to the tv set's balun or 75-ohm input jack.

   When a high-level distribution system (directional couplers) is used, it is highly unlikely that the head-end signal levels can be so high that the signals delivered to the tv sets will cause overload, even if a large antenna is used. In the extremely rare case where this *might* happen, select directional couplers with higher tap attenuation if the system is still in the design stage, or use attenuation for existing systems.

CHAPTER 10

Building the Antenna System

After the antenna and signal-distribution system have been designed, the next and final step is its physical construction. Before starting this, however, determine the best place to put the antenna. If you have a choice of locations (i.e., a very large flat roof) the proper approach is to locate the spot giving the strongest signal (suburban areas) or the most ghost-free picture (metropolitan areas). The test antenna is walked around the roof on the end of a 10-foot mast while someone observes the picture on a small portable tv set. Check all channels at each location. Keep one eye peeled for the roof edge, though! Do not try walking on even a flat roof with a big antenna or a long mast; it is very dangerous. Do not try this on a peaked roof no matter how small the antenna and mast are.

Never mount an antenna near power line. Never mount an antenna close to another antenna (except when stacking); the gain and pattern characteristics of each will be altered, and oscillator radiation from each one's tv set will be coupled to the other. Never mount an antenna close to metallic objects that are over a quarter wavelength at any frequency within the frequency range of the antenna. Such objects include radio or fm antennas, rain gutters, metal sheathing, air circulators, air conditioning equipment, and hatch covers. Never mount an antenna *less* than 7 feet from roof level on flat roofs; aside from the fact that this is too close for best performance, a person walking on the roof might run into an element end and lose an eye. Avoid chimney mounts if possible, since the output of an active chimney is rather corrosive.

MAST AND TOWER INSTALLATION

The erection of the antenna and its support may be done by the homeowner alone, or a small crew may be needed, depending on the type of roof and the size of the antenna(s) and its support. Flat roofs are easiest to work on; the antenna, mast, hardware, etc., can be carried up compactly and everything assembled up there. All of the recommended antennas are packed with instructions showing how to deploy the antenna elements, and on other matters important to proper installation and utilization. These instructions should be followed precisely to avoid damaging the antenna or obtaining substandard performance.

If the roof is peaked, as much work as possible should be done on the ground. This includes attaching the still-folded antenna to the mast, connecting the transmission line to the antenna, securing the transmission line to the mast in the appropriate manner, and attaching guy wires (if used). The antenna assembly is then hoisted up to the roof and inserted in the mount prepared for it. This requires two men: one to hoist from the roof and one to guide the mast end from a ladder next to the mount. If a guyed mast is used, more people may be needed to pull on the wires while the mast is set in place.

Tower installations are much more difficult because of their size and weight. The procedure differs radically according to the type of tower to be installed, so the manufacturer's instructions alone should be followed.

Never, never, erect a mast or tower of a length such that if it tips over it can contact power lines. Locate the antenna as far as possible from power lines. This is not only a good safety practice, it will also reduce the amount of electrical noise picked up by the antenna.

Antenna Connections

Unless the antenna has 75-ohm output provision, a balun is required for coaxial cable. Dress the balun leads so they clear the feeder lines (if any). If possible, provide a strain relief for the balun. This is easily accomplished on antennas where the underside of the boom is clear (all feeders on top) by taping the coax to the underside of the boom. A balun of the type shown in Fig. 6-2 is needed to permit this aproach.

Twinlead must be secured near the antenna terminals to avoid strain on the conductors. If a strain relief is not provided, mount a screw-thread standoff with the appropriate insert near the antenna terminals. Be certain to follow the instructions packed with the antenna regarding the preparation of the twinlead for connection. Some antennas require that the insulation be stripped from the line end, others have insulation-piercing terminals. For the latter type, the insulation *must* be left on, although oval and tubular foam lines will have to be pared down flat.

Anchoring Mounts

Nearly every type of mast mount except the chimney mount must be screwed to the structure on which the antenna is located. The type of fastener needed to securely install a mount depends on the material into which it will be sunk. For wooden walls, eaves, and shingle-covered roofs, the *lag screw* works well. These screws have square or hex heads, so they are turned by a wrench. When these are used to install roof or tripod mounts (as in Fig. 10-1), a pitch-soaked felt

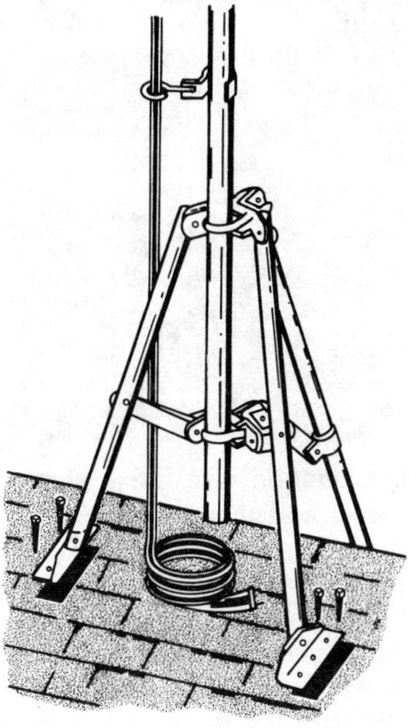

Fig. 10-1. Using lag screws and pitch pads to install a tripod mount on a shingle roof.

Courtesy Channel Master

pad should be placed between the mount and roof so rainwater won't leak through the screw holes. Pitch pads and lag screws are included with some tripods, such as those sold by Channel Master.

Fastening a mount to masonry (brick, cement, cinder block) is much more difficult. An oversized hole must be drilled in the proper spot and a device with soft, expandable material inserted in the hole. The mounting screw bites into this material and expands it to exert great pressure against the sides of the hole. A variety of devices are now available for this task (Fig. 10-2). The Rawlplug® at the extreme top of Fig. 10-2 is a lead sleeve surrounded by fiber; it is used in

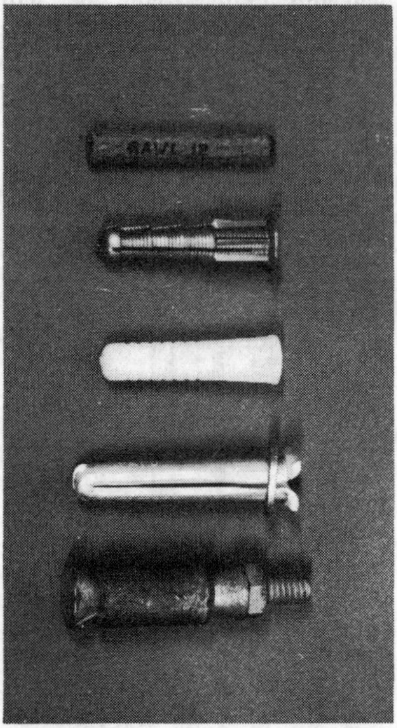

Fig. 10-2. Various masonry anchors.

conjunction with a lag screw. The next two are modern plastic versions, and they also are used with lag screws. The aluminum device near the bottom is used with a special drive-in nail, so it is good for securing guy wires. It can be used to mount cut-nail standoffs when you do not want to risk cracking the brick. The lead-anchor expansion bolt at the very bottom is self contained; after wrenching down the nut to anchor the device in the hole, the mast mount is slipped over the protruding bolt and fastened with another nut.

Guying Masts

Masts using base mounts, and many types of towers, require guying (i.e., holding the mast erect by means of taut wires). Even masts supported by wall or chimney mounts require guying if the mast height is over 15 feet.

Fig. 10-3 shows a typical guyed mast. While a 20-foot mast could get by with one set of guy wires, two sets are needed for the 30-foot mast in this illustration. The wires are attached to guy rings clamped over the joint of the interlocked mast sections. Although a set of guy wires fastened to the top of the mast is mechanically ideal, the wires will interfere with the antenna operation if placed this close. The

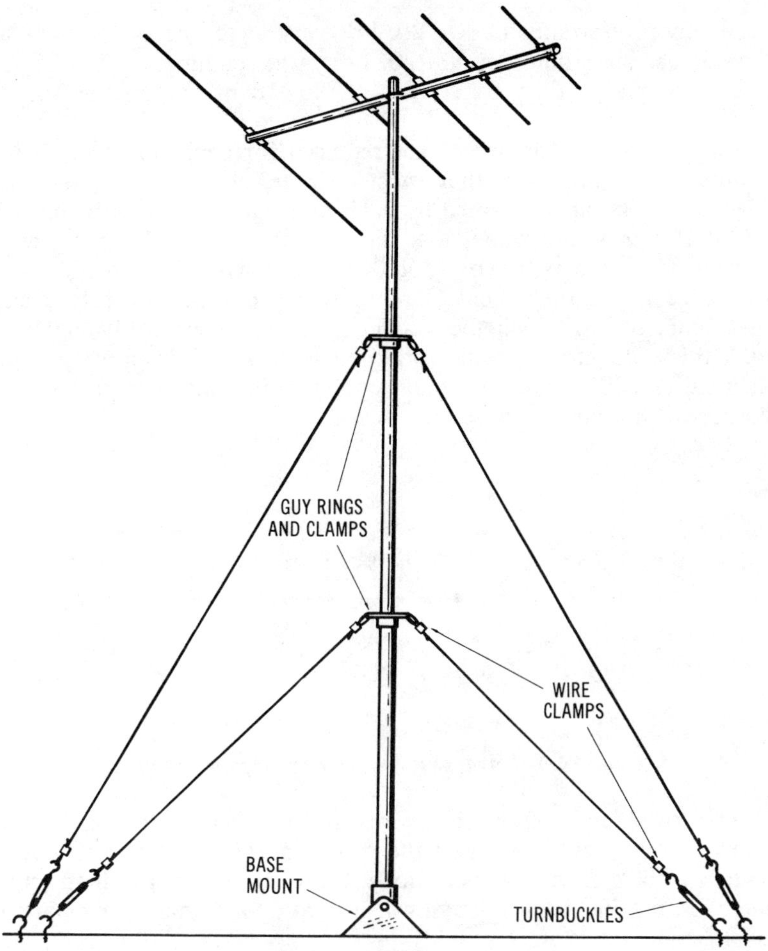

Fig. 10-3. 30-ft guyed mast.

one exception is if nonmetallic guying is used. Saxton 7000 is vinyl-covered fiberglass line a little more than ⅛ inch in diameter, which can be run close to the antenna elements without affecting antenna operation. A spot just under the antenna would be selected for the guy ring with this type of guy line. If a ready-made telescoping mast is used, the guy rings are already attached and properly located. The proper angle for each set of guy wires is somewhere between 30° and 60° relative to the mast. The larger the angle the less the stress on the guy wires produced by wind force.

If a small- to medium-sized vhf antenna or a uhf antenna is mounted atop a 20-foot mast, a relatively small guy wire can be used, say four strands of No. 20. For very large vhf and vhf-uhf antennas, use six strands of No. 20. For large antennas atop a 30-foot mast, no smaller than six strands of No. 18 or Saxton 7000 is advisable.

Either three or four wires are required for each set of guy wires. If three wires are used, they must be precisely spaced as shown in Fig. 10-4. The angle formed by each adjacent pair of wires must be 120°. If this is not done, the mast can be blown down by winds coming from the right (or should we say "wrong") direction. The four-wire guy arrangement is more work, but will tolerate greater misalignment. Although the system is strongest when all angles are 90°, this configuration will work well with a set of opposing angles as great as 120°. The rectangular configuration allows guying to each corner of the typical house.

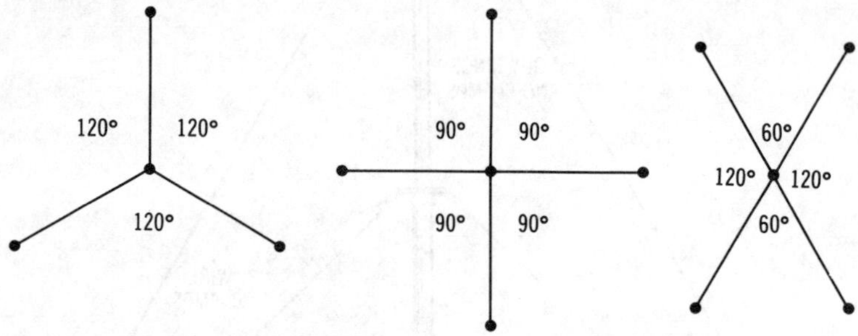

Fig. 10-4. Three- and four-wire guy arrangements.

The lower ends of the guy wires are attached to turnbuckles anchored by guy hooks set into the roof edge (Fig. 10-5) or sidewall. The guy hook anchorage is critical; it must be set in a firm structural member, not just in the shingles or insulating material. For maximum

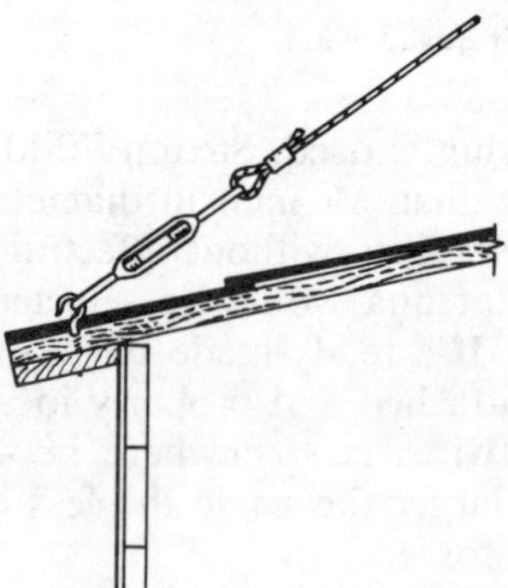

Fig. 10-5. Anchor and turnbuckle detail.

strength, drill a pilot hole through the shingles and insulation so the hook can be screwed into the structural member until just the bent part of the hook protrudes above the roof. Also, if the hook is set anywhere on the roof other than an edge (as shown in the illustration), roof cement must be smeared around the hook to prevent water seepage.

For relatively short masts bearing a uhf or small vhf antenna, the light-gauge guy wire can simply be looped through the guy ring or hook and its ends wrapped around the wire several turns to secure it. However, large masts bearing heavy antennas require much more care in securing the guy wires. The technique shown in Fig. 10-6, using a cable clamp to fasten the heavy-gauge guy wire, and a thimble to prevent wire breakage from a sharp bend, must be used at both the guy ring and turnbuckle ends of the guy wires.

The thimbles, guy wire, cable clamps, roof cement, hooks, turnbuckles, etc., needed for a first-class installation are available from many sources, among them Gavin, Lance, JFD, and Channel Master. The company that sells *all* of these fittings and materials, and has the largest variety of sizes, is IE Manufacturing.

When the base mounts and guy hooks have been installed, prepare the mast for erection by attaching the guy wires to the guy rings on the mast, and carefully tighten the cable clamps at this end. Cut each guy wire 2 feet longer than the calculated distance to be on the safe side. Thread the lower end of the guy wire through the turnbuckles and *lightly* fasten the cable clamps at this end. Open the turnbuckles to almost their full length. Temporarily erect the mast *without* the antenna; have one person hold it upright while you go to each turnbuckle and shorten the guy wires so each turnbuckle will just fit over its guy hook. Keep checking to see the mast remains vertical while doing this. When this is done lower the mast and attach the antenna.

Fig. 10-6. Guy wire attachment technique.

Raise the mast again and slip the turnbuckles over the hook anchors. Remove as much slack in the guy wires as possible by shortening the wire lengths. After the mast is truly vertical, fully tighten the cable clamps. Rotate the mast to orient the antenna for best reception, then tighten the base mount to prevent further rotation. Now begin to tighten the turn buckles to apply tension to the guy wires. Do this evenly, always watching to see that the mast is held perfectly vertical. If more than one set of guy wires is used, tension the lowest set first.

Grounding the Mast

Grounding an antenna mast accomplishes two things; it prevents static charges from accumulating on the antenna (very important in dry climates) and it helps protect against damage caused by lightning strokes. Lightning effects and the hardware used to prevent

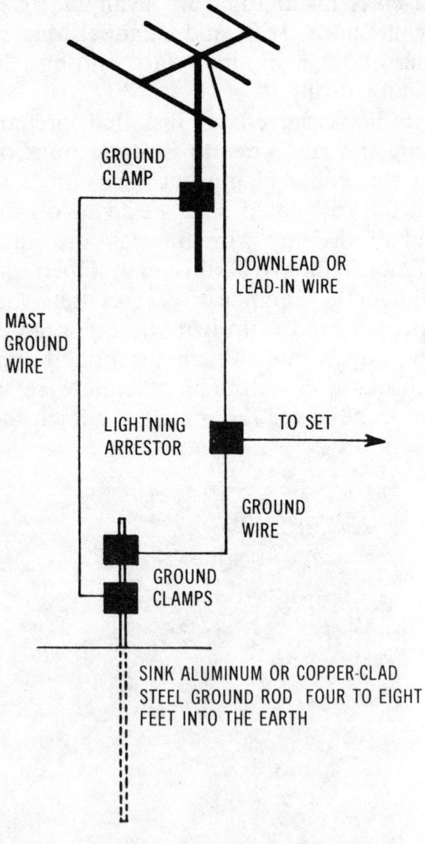

Courtesy Winegard Co.

Fig. 10-7. Ground system.

damage were discussed in Chapter 7; installation details are given here.

Number 8 aluminum wire should be attached to the bottom of the antenna mast with a ground clamp, or it can be bolted to the mast if you don't mind drilling an extra hole. Run this wire down the side of the house to a ground rod driven into the earth (Fig. 10-7) or a length of brass pipe buried underground. Sufficient depth is needed in either case to reach moist soil. Connect another length of No. 8 wire from the ground rod to a grounding block (for coax) or lightning arrestor (unshielded twinlead) located at the point where the transmission line enters the structure. The grounding block is also a convenient point at which to change (if you so desire) from a large-diameter coax used for a long run to the antenna, to a more manageable RG59 size for routing indoors. For shielded twinlead you will have to improvise; try carefully removing the vinyl outer jacket to expose the shield and drain wires; then attach the ground wire with a homemade clamp of sheet aluminum and smear epoxy over the spot to protect and strengthen the line.

For maximum protection the ground wires and transmission line should be held away from a wooden wall or roof by standoffs. A direct stroke might melt the ground wires; so the spacing will prevent scorch marks on the wood.

Grounding the mast also grounds the antenna in most cases. If the antenna design is such that an insulating mount is used between the antenna and mast, set the antenna about 1 foot down from the top of the mast so lightning strokes will hit the mast rather than the antenna.

INSTALLING THE TRANSMISSION LINE

Techniques for handling the many small problems and details involved in properly getting the transmission line from the antenna to the tv set or signal-distribution system are covered in this section.

Installing F-Connectors

Often problems such as intermittent signal or very weak signal are due to poorly installed F-connectors. The problem may be due to failure to prepare the cable end properly or a crumpled foil shield. Both of these faults, and most others, can be avoided by simply paying attention to what you are doing. The Jerrold Applications Bulletin reproduced in Fig. 10-8 describes in detail how to install F-connectors on foil-and-braid coax. These instructions are correct for *any* size coax.

Step 9 in the instructions mentions crimping the ferrule (crimp ring) to secure the F-connector the cable. To be certain of good

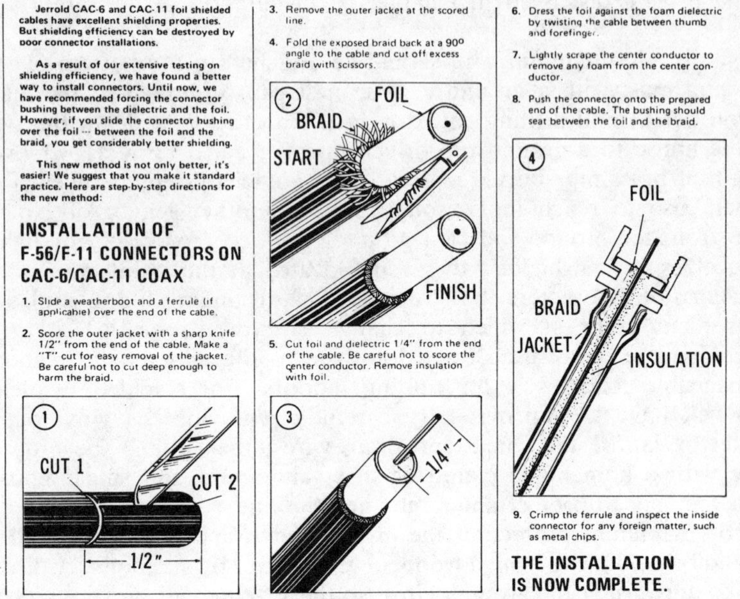

Jerrold CAC-6 and CAC-11 foil shielded cables have excellent shielding properties. But shielding efficiency can be destroyed by poor connector installations.

As a result of our extensive testing on shielding efficiency, we have found a better way to install connectors. Until now, we have recommended forcing the connector bushing between the dielectric and the foil. However, if you slide the connector bushing over the foil --- between the foil and the braid, you get considerably better shielding.

This new method is not only better, it's easier! We suggest that you make it standard practice. Here are step-by-step directions for the new method:

INSTALLATION OF F-56/F-11 CONNECTORS ON CAC-6/CAC-11 COAX

1. Slide a weatherboot and a ferrule (if applicable) over the end of the cable.
2. Score the outer jacket with a sharp knife 1/2" from the end of the cable. Make a "T" cut for easy removal of the jacket. Be careful not to cut deep enough to harm the braid.
3. Remove the outer jacket at the scored line.
4. Fold the exposed braid back at a 90° angle to the cable and cut off excess braid with scissors.
5. Cut foil and dielectric 1/4" from the end of the cable. Be careful not to score the center conductor. Remove insulation with foil.
6. Dress the foil against the foam dielectric by twisting the cable between thumb and forefinger.
7. Lightly scrape the center conductor to remove any foam from the center conductor.
8. Push the connector onto the prepared end of the cable. The bushing should seat between the foil and the braid.
9. Crimp the ferrule and inspect the inside connector for any foreign matter, such as metal chips.

THE INSTALLATION IS NOW COMPLETE.

Courtesy TACO/Jerrold

Fig. 10-8. Instructions for attaching F-connectors to foil-and-braid shield cables.

results, a crimping tool made for this purpose is needed. Ordinary linesman's pliers can be used with some steel ferrules, but considerable time and skill are required to get an acceptable crimp. Low-cost crimping tools are available from chainstores like Lafayette Radio for under $5, or a professional-grade parallel-jaw crimper like the RMS Electronics 1885 shown in Fig. 10-9 can be purchased. Tools similar to this one are sold by several suppliers, including Winegard, Channel Master, Blonder-Tongue, and Jerrold.

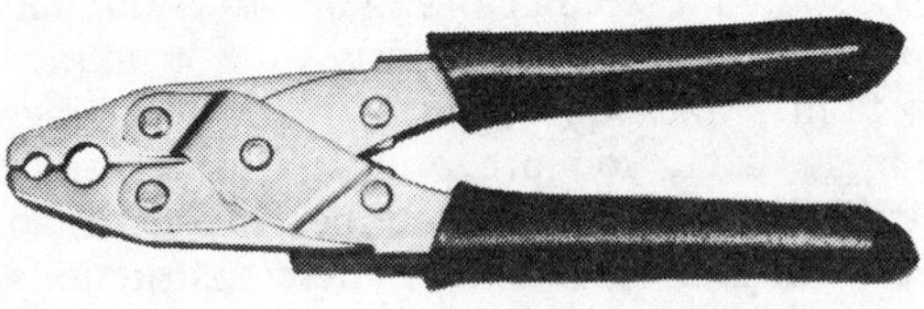

Fig. 10-9. RMS Electronics 1885 crimping tool.

Waterproofing Cable Connections

For reliable connections over a period of many years, all outdoor cable connections should be protected from the weather. For simple antenna systems, this may involve only the balun connection, which is usually well protected with the rubber boot supplied with the balun. For a complicated antenna system many spots must be protected—rotator link connections, band combiners, signal combiners, grounding blocks, etc. Many of these devices come equipped with rubber boots to cover the mated F-connectors, but seepage can still occur because the shape of the F-connector will not permit a watertight seal by a rubber boot. Catv-system installers improve the seal by applying a plastic or rubber sealant around the joint to augment the rubber boot. For home and small matv systems good results are obtainable by carefully wrapping the joint with ¼-inch-wide plastic electrical tape. When applied in mild or warm weather (the best time to do antenna work), this tape will stretch under tension and conform to every contour of the mated F-connectors, even the threads. After thoroughly wrapping the joint (from the body of the device to the cable itself), smear a little quick-drying glue (Duco® cement, for example) on the tape, particularly at the highest point of the joint. Cable connections so treated by the author were found to be in pristine condition when checked years later.

Rotator Links

A highly flexible piece of RG59-sized foam coax having a *copper-braid* shield should be used as the section of transmission line bridging the rotated and fixed mast sections. This type of coax can better stand the stress of continued flexion than foil-shield coax. Belden 8221 is recommended for this application.

When at least 2 feet of rotated mast section is avilable between the rotator and antenna, the technique shown in Fig. 10-10 is ideal. Rotate the antenna to the extreme north limit, *approaching from the east.* Then wind the coax snugly around the rotated mast two turns in the direction shown. This way the rotator link will unwind as the antenna moves N→E→S→W→N, and rewind as it rotates N→W→S→E→N again. If you wind the link the wrong way, it will either break or halt the antenna rotation.

Fastening the Line

Coax and shielded twinlead are the easiest lines to install. They are simply taped to the antenna mast or tower with 1-inch-wide plastic electrical tape, and run along the junction of the roof and parapet wall to where the line begins its downward run to the building entry. If the run is over 10 feet, either type of line must be secured

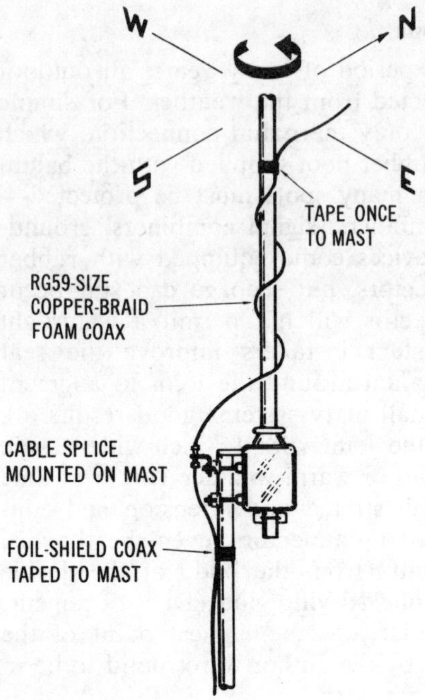

Fig. 10-10. Installation technique for rotator link.

to prevent its abrasion by the building face or edge of the parapet wall when wind blown. If a drain downspout is nearby, the line can be taped to it. If not, an appropriate type of standoff should be used to secure the line to the building every 10 feet or so. Use standoffs having vhf inserts for RG59-sized coax and rotator wire, uhf inserts for RG6-sized coax and small-sized shielded twinlead (Belden 9090), and special oval inserts for large-sized shielded twinlead (Belden 8290). See Chapter 7 for data on types of standoffs and the inserts mentioned.

Unshielded twinlead requires a great deal of care and hard work to produce an acceptable installation. The easiest part is the line run down the mast; the only point to remember here is to use 7-inch screw-thread standoffs in a strap mount for weak-signal areas, rather than the shorter snap-on style. The hard part starts when the line run changes direction; two standoffs are needed for each 90° turn (Fig. 10-11A). The turns should be gradual; no sharp bends are advisable. The real difficulty is getting the line past metal rain gutters; none of the insulators commercially available are suitable for fastening to metal gutters, and none are long enough to screw into the roof and hold the line away if the gutter is sizable. You can try bending the end of a 7-inch screw-thread standoff into a

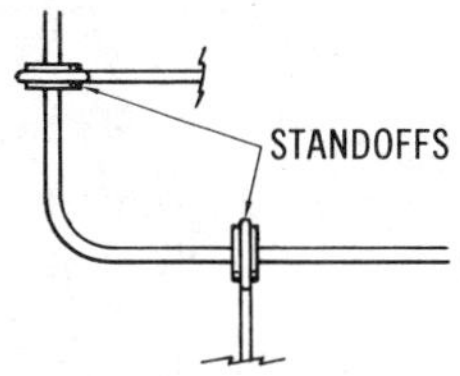

(*A*) *90° turn of line.*

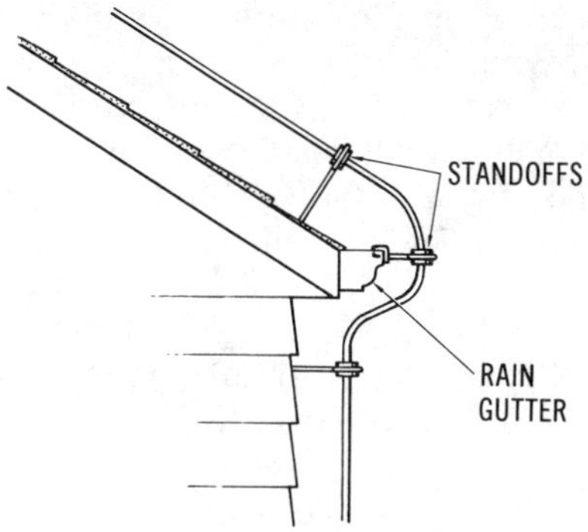

(*B*) *Running line past gutter.*

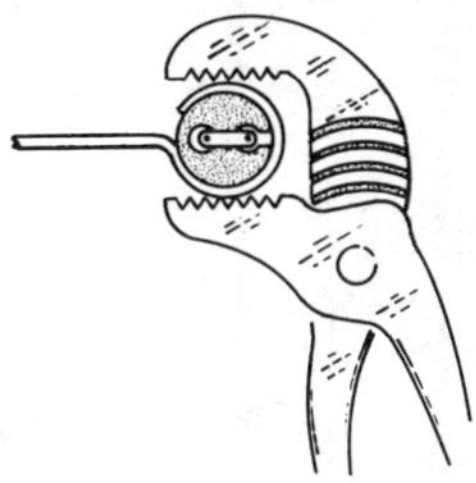

(*C*) *Installing in common standoffs.*

Fig. 10-11. Installation techniques for unshielded twinlead.

loop and bolting this to the outside of the gutter, as shown in Fig. 10-11B.

The same rain-gutter downspout that made a convenient support for coax and shielded twinlead downruns should be avoided like the plague when unshielded twinlead is used. Also avoid other transmission lines (of any kind), rotator cable, telephone wires, power lines, fire escapes, and aluminum siding. In short, never run unshielded parallel to any wires, or close to any large metal surfaces or structures. To minimize pickup from other tv transmission lines and power lines, twist the twinlead about one-half turn per foot for long straight runs. This will minimize the unbalancing effect of nearby metallic objects.

Use sufficient standoffs to keep the line from whipping in the wind, as well as hold the line away from everything. Space the standoffs at irregular intervals to avoid problems at a particular frequency. When installing twinlead in common standoffs (those having a metal ring around the polyethylene insert), orient the insert as shown in Fig. 10-11C before squeezing the ring with Channellock® pliers. By rotating the insert so the plane of the conductors is perpendicular to the direction of force, the conductors of the line are not squeezed together, and an impedance irregularity is avoided. Also notice that the opening in the insert is on the inside, a precaution against the line working loose. Use standoffs with inserts appropriate to the type of twinlead used: IE Manufacturing oval inserts for oval foam-encased twinlead (Belden 8285), and either IE Manufacturing Line-Lok® or Lafayette-style standoffs for tubular twinlead (Belden 8275). See Chapter 7 for details on these standoffs.

Building Entry

Coax entry is simple; bore a hole through a convenient wall as shown in Fig. 10-12A. The hole must slope in the direction shown

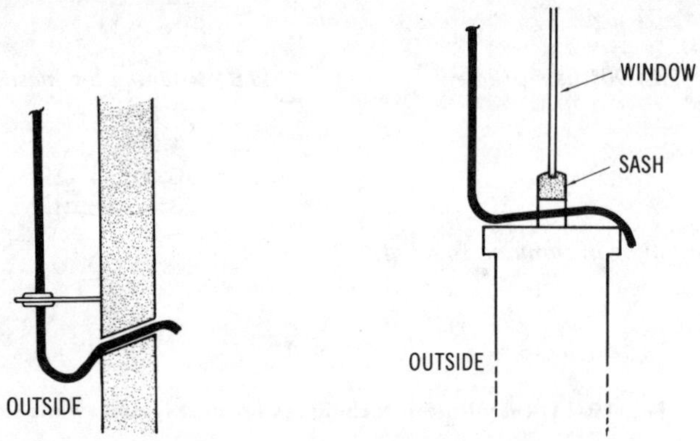

(*A*) *Through a wall.* (*B*) *Through a window.*

Fig. 10-12. Entry techniques for coaxial cable.

so water will not run into the house. (The "drip loop" in the coax is also done for this purpose.) If the wall is solid, the coax can be pushed through by itself. If the wall has airspace, an aluminum or stiff plastic tube must be installed first, or the coax will miss the inside hole.

If drilling through a wall is too difficult or not permitted, building entry is easily accomplished through a window. Place a piece of wood about the same width as the window sash between the sash and window sill (Fig. 10-12B). The height of the block need only be enough to permit drilling a sloping hole for the coax. Drill the hole at one end of the wooden insert so the cable can be secured with a standoff or cable clip on the adjacent wall.

Twinlead installations require more than just the drip loop and sloping entry hole. A special large-diameter tube for this purpose should be installed in the hole. The tube is designed so the section of twinlead passing through it is tautly held, equidistant from the sides of the tube. This minimizes the impedance disruption and unbalance caused by the building materials. The Radio Shack 15-1200 is a good example of this device. However, if the building has aluminum siding or drilling is not permitted, the entry technique shown in Fig. 10-13 can be used. In this case a window insert made of wood or plastic slats (Fig. 10-13A) should be used. Also, if aluminum storm windows are used, a high, flat strip of plastic should be placed in the aluminum window track to keep the line away from the metal frame and sash (Fig. 10-13B).

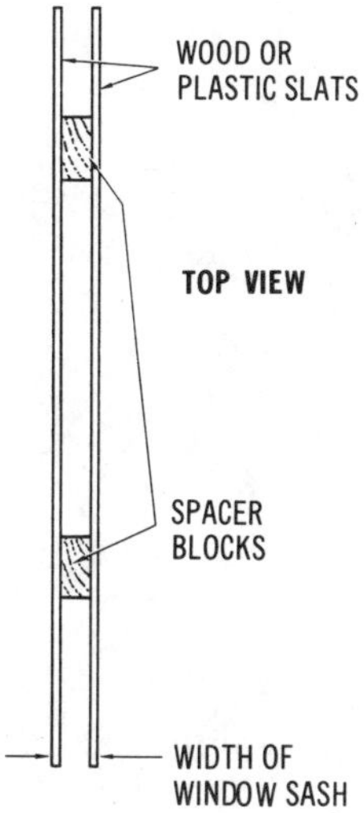

(*A*) *Hollow insert.*

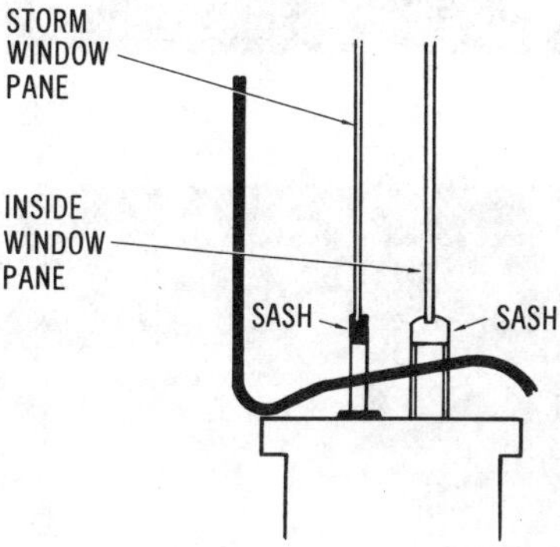

(*B*) *Technique for storm window.*

Fig. 10-13. Window entry techniques for unshielded twinlead.

Once inside the house, unshielded twinlead should be cut to just a couple of feet longer than the minimum length needed to reach the tv set or amplifier. Never coil excess twinlead.

DISTRIBUTING THE SIGNAL

Multiset signal distribution is always by means of coaxial cable in quality installations. The big decision is whether an in-the-wall or outside-the-wall route is used for the coax.

In-Wall Routing

For houses under construction, or buildings with false ceilings, in-the-wall (or in-the-ceiling) routing is obviously best. Entry to the room is made by running a short length of coax called a *drop*

(*A*) *With F-81 cable splice.*

(*B*) *With directly mounted directional coupler.*

Fig. 10-14. Wall plate mounts.

cable from the tap port of the directional coupler to an F-81 cable splice mounted in a wall plate (Fig. 10-14A). If the wall board or panel thickness is less than ⅜ inch, an RMS Electronics CA-1090 series directional coupler can be mounted directly on the wall plate (Fig. 10-4B) and the drop cable and splice eliminated. The Jerrold DFT-style directional couplers can be used in this manner with wall-board up to ¾ inch thick. Thicker paneling and/or directional styles require the drop-cable-and-splice technique.

The wall plate can be fastened to the wall (or baseboard) in one of two ways. The easiest, which is suitable for wooden baseboards or paneling, is to simply cut a hole big enough to accommodate the directional coupler used, and fasten the wall plate with wood screws. Wood screws usually do not hold well enough in plasterboard, however, so wall-plate brackets are advisable. Fig. 10-15 shows how RMS Electronics Model CA-5035 wall-plate brackets are installed.

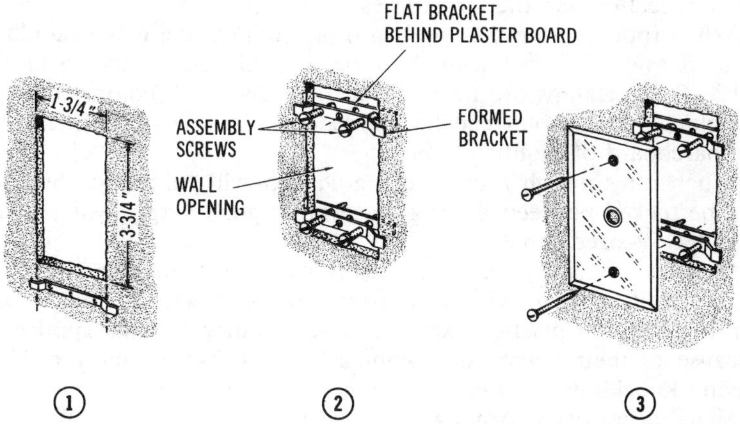

Courtesy RMS Electronics

Fig. 10-15. Steps involved in installing wall-plate brackets.

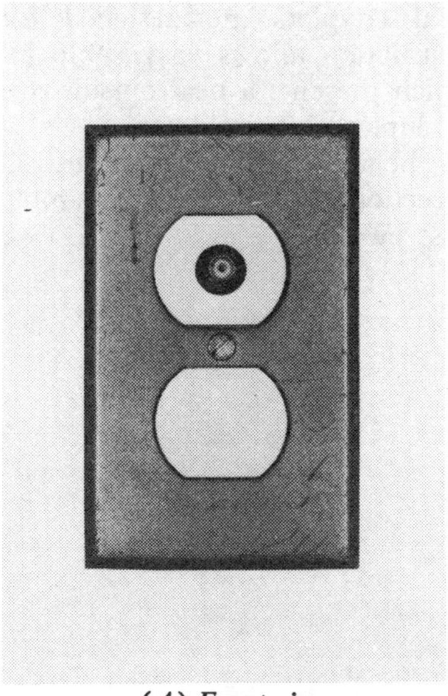

(*A*) *Front view.*

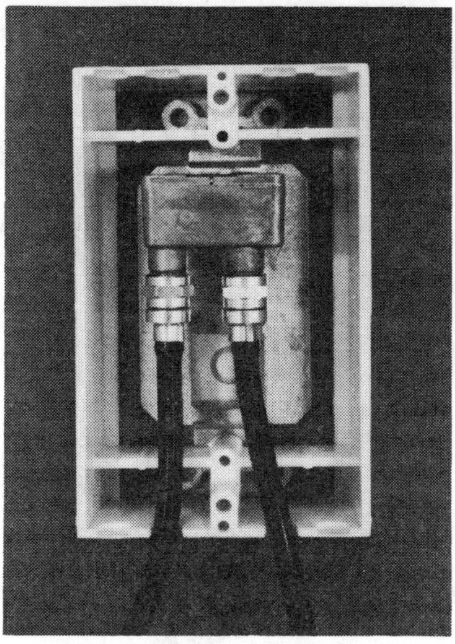

(*B*) *Rear view.*

Fig. 10-16. Flush-mount wall tap mounted in a plastic outlet box.

External Routing

Installing cable in the walls of existing buildings, particularly homes, is possible but very expensive. A cheaper and much quicker alternative is to route the cable external to the wall, either along the baseboard or molding, or at the junction of floor and wall. Cable that is RG59-sized is preferred for this kind of installation, as it is less noticeable than the larger sizes.

When routed along the baseboard or molding, the cable should be secured every few feet with the type of cable clamp shown in Fig. 7-12. These clamps are available in light and dark colors, although the clamps and cables can be painted over afterward to blend into the baseboard or molding.

When run through rooms having wall-to-wall carpeting, the cable can be tucked between the rug and wall; a good carpet will virtually bury RG59-sized cable.

Outlet boxes are required to neatly mount the directional coupler on the wall board in visible locations. Metal electrical outlet boxes can be used, but plastic boxes are more common for this application because of their lower cost. Standard plastic boxes, many of them having knockouts for the cable, are sold by most manufacturers of tv signal accessories. Among them are Jerrold (UTSH-2), Winegard (SM-1), Channel Master (7027), and Antenna Corp. of America (ELT-600).

The Jerrold DFT-series directional couplers are designed for mounting in outlet boxes. Their construction allows easy cable installation, as shown in Fig. 10-16, which presents a neat outside appearance very much like the standard duplex electric outlet.

For "hidden" outlet locations (e.g., behind drapery) or industrial buildings, the RMS CA-1090 style directional coupler can be simply fastened to the baseboard with a wood screw.

APPENDIX **A**

Abbreviations Used in This Book

af array factor
atten attenuation, attenuator
BW bandwidth, beamwidth
catv community antenna tv
CB citizens band
CH, chan channel
coax coaxial cable
D diameter
dB decibel
dBmV dB relative to 1 millivolt
Dlr dealer
f frequency
fbr front-to-back ratio
fm frequency modulation
fsr front-to-side (ratio)
ft feet, foot
H/L high/low (band vhf)
in inch(es)
kg kilogram(s)
L length
lbs pounds
matv master antenna tv
MHz megahertz
mV millivolt(s)
No. number
OD outside diameter
rf radio frequency
rfi radio-frequency interference
rfs relative field strength
snr signal-to-noise ratio
swr standing-wave ratio
tv television
uhf ultrahigh frequency
V volt(s)
vhf very high frequency
vhf-H vhf high band
vhf-L vhf low band
Wt weight
μ**V** microvolt(s)
$^\circ$ degrees
λ wavelength
Ω ohms
$<$ angle

APPENDIX **B**

Directory of Manufacturers

Contact the manufacturers at the address listed below for sales information if their products are not available from your local distributor.

Antenna Corp. of America
Box 865
Burlington, IA 52601

Antennacraft Co.
Box 1005
Burlington, IA 52601

AVA Electronics Corp.
242 Pembroke Ave.
Lansdowne, PA 19050

Belden Corp.
Box 1331
Richmond, IN 47374

Blonder-Tongue Laboratories
1 Jake Brown Road
Old Bridge, NJ 08857

Channel Master
Route 209
Ellenville, NY 12428

Finney Co.
34 W. Interstate St.
Bedford, OH 44146

Gavin Electronics
1450 U.S. Route 22
Somerville, NJ 08876

IE Manufacturing
933 E. Remington Rd.
Schaumburg, IL 60195

JFD Electronics Corp.
Pine Tree Road
Oxford, NC 27565

Lance Industries
Box 4156
Sylmar, CA 91342

Radio Shack
1 Tandy Center
Ft. Worth, TX 76102

RCA Distributor & Special Products
Antenna Merchandising
2000 Clements Bridge Rd.
Deptford, NJ 08096

RMS Electronics
50 Antin Place
Bronx, NY 10462

Saxton Products
215 N. Route 303
Congers, NY 10920

SITCO Antennas
Box 20456
Portland, OR 97220

TACO/Jerrold
1 Taco St.
Sherburne, NY 13460

Winegard Co.
3000 Kirkwood St.
Burlington, IA 52601

Index

A

Abbreviations used in this book, 245
All-channel antennas, 85-104
  commercial, 88-104
    measured performance comparison, 101-104
    medium-signal, 94-100
    strong- to medium-signal, 89-94
    weak-signal, 100-101
  isolation networks, 86-87
Amplifiers, 133-144
  distribution, 143-144
  preamplifiers, 136-142
    indoor, 141-142
    mast-mounted, 136-141
Anchoring mounts, mast and tower installation, 228-235
Antenna
  connections, mast and tower installation, 228-229
  hardware, 151-167
    guying, 158-160
    lightning-protection devices, 163
    masts and towers, 151-154
    mounting brackets and devices, 154-157
      base mounts, 155
      chimney mounts, 155-156
      mounting kits, 157
      tripod mounts, 154-155
      wall mounts, 156-157

Antenna—cont
  principles, tv, 17-28
    bandwidth, 22-24
    directionality, 20-22
    frequency and wavelength, 17-18
    gain, 24-25
    impedance, 19-20
    parasitic antennas, 25-28
    resonance, 18-19
  rotators, 163-167
  selection, head-end design, 207-209
  standoffs, 161-162
  system basics, 7-28
    decibel notation, 11-12
    picture quality, 12-16
    reception range, 16-17
    requirements of tv, 7
    signal level systems, 10-11
    tv
      antenna principles, 17-28
      bands and channel assignments, 8-9
  system building, 227-244
    distributing the signal, 241-244
      external routing, 244
      in-wall routing, 241-243
    installing the transmission line, 235-241
      building entry, 240-241
      fastening the line, 237-240

Antenna—cont
system building
installing the transmission line
installing F-connectors, 235-237
rotator links, 237
mast and tower installation, 228-235
anchoring mounts, 229-230
antenna connections, 228-229
grounding the mast, 234-235
guying masts, 230-234
system designing, 203-224
distribution-system design, 213-224
calculating tap values, 217-224
receiver interface, 224
system layout, 214-216
head-end design, 206-213
antenna selection, 207-209
preamplification, 211-213
transmission-line selection, 207-209
matching distribution system to head end, 225-226
signal survey, 203-206
attenuator technique, 204-205
field-strength-meter technique, 204
Attenuation, 105
Attenuators, 144-146
attenuator pads, 144-145
tilt, 145-146
Attenuator technique, signal survey, 204-205

B

Back matching, 132
Balanced line, 106-110
foam twinlead, 108-109
open line, 106
shielded twinlead, 109-110
twinlead, 107-108
Baluns, 105, 118-120
Band separators, 125-130
Bands and channel assignments, tv, 8-9
Bandwidth, antenna, 22-24
Base mounts, antenna, 155
Beamwidth, 22
Bimoding, 36
Bowtie and reflector uhf antennas, 61-66
4-bay bowtie, 62-66
single-bay bowtie, 62
Broadband element, 23
Building
entry, transmission line, 240-241
the antenna system, 227-244
distributing the signal, 241-244
external routing, 244
in-wall routing, 241-243
installing the transmission line, 235-241
building entry, 240-241
fastening the line, 237-240
installing F-connectors, 235-237
rotator links, 237
mast and tower installation, 228-235
anchoring mounts, 229-230
antenna connections, 228-229
grounding the mast, 234-235
guying masts, 230-234

C

Cable
coaxial, 110-113
splice, 149
switch, 148-149
Calculating tap values, distribution-system design, 217-224
Channel
assignments, 8-9
-mixing networks, 198-202
Chimney mounts, antenna, 155-156
Coax, *see* Coaxial cable
Coaxial cable, 110-113
Combiners, signal, 124-125
Combining antennas, rules for, 175-177
Commercial
all-channel antennas, 88-104
measured performance comparison, 101-104
medium-signal, 94-100
strong- to medium-signal, 89-94

Commercial—cont
all-channel antennas
weak-signal, 100-101
corner-reflector uhf antennas, 67-69
single-band antennas, 34-35
uhf multiband antennas, 41-60
medium
-signal antennas, 44-50
to weak-signal antennas, 50-53
strong-signal antennas, 42-44
weak-signal antennas, 53-58
Comparisons
all-channel antenna performance, 101-104
transmission lines, 113-115
uhf antenna performance, 79-83
vhf antenna performance, 58-60
Corner-reflector uhf antennas, 66-74
commercial, 67-69
parabolic-cylinder, 67
yagi, 69-74
Couplers
directional, 131-133
yagi, 196-198

D

dBmV, *see* Decibel system
Decibel
notation, 11-12
system, 10-11
Designing the antenna system, 203-226
distribution-system design, 213-224
calculating tap values, 217-224
receiver interface, 224
system layout, 214-216
head-end design, 206-213
antenna selection, 207-209
preamplification, 211-213
transmission-line selection, 210-211
matching the distribution system to the head end, 225-226
signal survey, 203-206
attenuator technique, 204-205
field-strength-meter technique, 204
Dipole, 19
Directional couplers, 131-133
Director
arrays, accessory, increased signal output with, 169-180
definition, 25
Directory of manufacturers, 247
Distributing the signal, antenna system, 241-244
external routing, 244
in-wall routing, 241-244
Distribution
amplifiers, 143-144
-system design, 213-224
calculating tap values, 217-224
receiver interface, 224
system layout, 214-216
Driven element, 25
Drop
cables, 131
tap, 132

E

8-bays, 193
Extension stubs, 38
External routing, signal distributing, 244

F

Fastening the transmission line, 237-240
F-connectors
description, 146-148
installing, 235-237
Field-strength-meter signal survey, 204
Filters, 193-196
Fm reception on multiband antennas, 39-41
Foam twinlead, 108-109
Folded dipole, 24
4-bay uhf bowtie antenna, 62-66
Frequency and wavelength, antenna, 17-18
Front-to-side ratio, 21

G

Gain, antenna, 24-25

Ghost and interference elimination, 180-196
filters, 193-196
high-performance antennas, 181-183
horizontal stacking, 187-193
stagger stacking, 185-186
vertical stacking, 183-185
Ghosts, 12-15
Grade of tv picture, 15
Grounding the mast, 234-235
Guying
hardware, antenna, 158-160
masts, 230-234

H

Half-power points, 21-22
Hardware, antenna, 151-167
antenna rotators, 163-167
guying hardware, 158-160
lightning-protection devices, 163
masts and towers, 151-154
mounting brackets and devices, 154-157
base mounts, 155
chimney mounts, 155-156
mounting kits, 157
tripod mounts, 154-155
wall mounts, 156-157
standoffs, 161-162
Head-end design, 206-213
antenna selection, 207-209
preamplification, 211-213
transmission-line selection, 210-211
Head ends, matv, 196-202
channel-mixing networks, 198-202
yagi couplers, 196-198
High
/low director, 39
-performance antennas, 181-183
Higher antennas, increased signal output, 173-174
Horizontal stacking, 187-193
Hoverman uhf antenna, 76-78

I

Impedance, antenna, 19-20
Increased signal output, 169-180
accessory director arrays, 170-173
higher antennas, 173-174
larger antenna, 169-170
preamplifiers, 179-180
separate antennas, 174-177
stacked antennas, 177-179
Indoor
preamplifiers, 141-142
uhf antenna, 62
Installing
F-connectors, 235-237
transmission line, 235-241
building entry, 240-241
fastening the line, 237-240
installing F-connectors, 235-237
rotator links, 237
Interference and ghost elimination, 180-196
filters, 193-196
high-performance antennas, 181-183
horizontal stacking, 187-193
stagger stacking, 185-186
vertical stacking, 183-185
Insertion loss, 118
In-wall routing, signal distributing, 241-244
Isolation networks, all-channel antennas, 86-87

L

Larger antenna, increased signal output with, 169-170
Line
balanced, 106-110
open, 106
transmission, 105-115
Lobe, 21
Log-periodic
principle, 33-34
uhf antennas, 74-76

M

Manufacturers, directory of, 247

Mast
and tower installation, 228-235
anchoring mounts, 229-230
antenna connections, 228-229
grounding the mast, 234-235
guying masts, 230-234
-mounted preamplifiers, 136-141
Masts and towers, antenna hardware, 151-154
Matching distribution system to head end, 225-226
Matv head ends, 196-202
channel-mixing networks, 198-202
yagi couplers, 196-198
Measured performance comparison
all-channel antennas, 101-104
uhf antennas, 79-83
vhf antennas, 58-60
Medium
-signal
all-channel antennas, 94-100
vhf multiband antennas, 44-50
to weak-signal vhf multiband antennas, 50-53
Microvolt, 10
Mid band, 9
Millivolt, 10
Mounting
brackets and devices, antenna, 154-157
base mounts, 155
chimney mounts, 155-156
mounting kits, 157
tripod mounts, 154-155
wall mounts, 156-157
kits, antenna, 157
Multiband vhf antennas, 35-41
fm reception, 39-41
multimode directors, 39
multimoding techniques, 36-39
Multimode directors, vhf, 39
Multimoding techniques, vhf, 36-39

N

Narrow-band
antennas, 22
yagi, corner-reflector, 73-74
Networks
channel-mixing, 198-202
isolation, all-channel antennas, 86-87
Noise figure, preamplifier, 133

O

Open line, 106

P

Pads, attenuator, 144-145
Parabolic
-cylinder corner-reflector uhf antennas, 67
-reflector uhf antennas, 78-79
Parasitic antennas, 25-28
Performance comparison
all-channel antennas, 101-104
uhf antennas, 79-83
vhf antennas, 58-60
Picture quality, 12-16
ghosts, 12-15
signal-to-noise ratio, 15-16
Preamplification, head-end design, 211-213
Preamplifiers, 136-142, 179-180
increased signal output with, 179-180
indoor, 141-142
mast-mounted, 136-141

Q

Quad stacks, 191

R

Range, tv antenna reception, 16-17
Receiver interface, distribution-system design, 213-224
Reception range, antenna, 16-17
Reflector
and bowtie antennas, 61-66
4-bay bowtie, 62-66
single-bay bowtie, 62
definition, 25

Relative field strength, array, 188
Requirements of a tv antenna system, 7
Resonance, antenna, 18-19
Right-angle fitting, 149-150
Rotator links for transmission line, 237
Rotators, antenna, 163-167
Routing of signal, 241-244
  external, 244
  in-wall, 241-244

S

Selection criteria, antenna, 174-175
Separate antennas, increased signal output with, 174-177
Separators, band, 125-130
75-ohm signal splitters, 122-124
Shield, coaxial cable, 110
Shielded twinlead, 109-110
Signal
  combiners, 124-125
  distributing, 241-244
    external routing, 244
    in-wall routing, 241-244
  -distribution components, 117-150
    amplifiers, 133-144
      distribution amplifiers, 143-144
      indoor preamplifiers, 141-142
      mast-mounted preamplifiers, 136-141
    attenuators, 144-146
      attenuator pads, 144-145
      tilt, 145-146
    baluns, 118-120
    band separators, 125-130
    directional couplers, 131-133
    miscellaneous, 146-150
      cable splice, 149
      cable switch, 148-149
      F-connectors, 146-148
      right-angle fitting, 149-150
      terminators, 148
    signal splitters, 120-125
      75-ohm, 124
      signal combiners, 124-125
      300-ohm, 122-124
Signal—cont
  level systems, 10-11
    decibel, 10-11
    voltage, 10
  optimization techniques, 169-202
    increased signal output, 169-180
      accessory director arrays, 170-173
      higher antennas, 173-174
      larger antenna, 169-170
      preamplifiers, 179-180
      separate antennas, 174-177
      stacked antennas, 177-179
    ghost and interference elimination, 180-196
      filters, 193-196
      high-performance antennas, 181-183
      horizontal stacking, 187-193
      stagger stacking, 185-186
      vertical stacking, 183-185
    Matv head ends, 196-202
      channel-mixing networks, 198-202
      yagi couplers, 196-198
  splitters, 120-125
    75-ohm, 124
    signal combiners, 124-125
    300-ohm, 122-124
  survey, 203-206
    attenuator technique, 204-205
    field-strength-meter technique, 204
  to-noise ratio, 15-16
Single
  -band antennas, 32-35
    commercial, 34-35
    log-periodic principle, 33-34
  -bay uhf bowtie antenna, 62
  -channel vhf antennas, 29-32
Splice, cable, 149
Splitters, signal, 120-125
  75-ohm, 124
  signal combiners, 124-125
  300-ohm, 122-124
Stacked antennas, increased signal output with, 177-179
Stacking antennas, 183-193
  horizontal, 187-193

Stacking antennas—cont
  stagger, 185-186
  vertical, 183-185
Stagger stacking, 185-186
Standing-wave ratio, 118
Standoffs, 161-162
Steps to designing antenna system, 203
Strong
  -signal vhf multiband antennas, 42-44
  to medium-signal all-channel antennas, 89-94
Super band, 9
Switch, cable, 148-149
System layout, distribution-system design, 213-224

T

Tap
  directional coupler, 132
  values, calculating, 217-224
Terminators, 148
300-ohm signal splitters, 122-124
Tilt attenuators, 145-146
Tower and mast installation, 228-235
  anchoring mounts, 229-230
  antenna connections, 228-229
  grounding the mast, 234-235
  guying masts, 230-234
Towers and masts, antenna hardware, 151-154
Translator channels, antenna for, 64
Transmission lines
  balanced line, 106-110
    foam twinlead, 108-109
    open line, 106
    shielded twinlead, 109-110
    twinlead, 107-108
  coaxial cable, 110-113
  comparison of, 113-115
  installing, 235-241
    building entry, 240-241
    fastening the line, 237-240
    installing F-connectors, 235-237
    rotator links, 237
  selection, head-end design, 210-211
Tripod mounts, antenna, 154-155
Trunk line, 131
TV
  antenna
    principles, 17-18
      bandwidth, 22-24
      directionality, 20-22
      frequency and wavelength, 17-18
      gain, 24-25
      impedance, 19-20
      parasitic antennas, 25-28
      resonance, 18-19
    system requirements, 7
  bands and channel assignments, 8-9
Twinlead, 107-108

U

Uhf
  antennas, 61-83
    bowtie and reflector, 61-66
      4-bay bowtie, 62-66
      single-bay bowtie, 62
    corner-reflector, 66-74
      commercial, 67-69
      parabolic-cylinder, 67
      yagi, 69-74
    Hoverman, 76-78
    log-periodic, 74-76
    parabolic-reflector, 78-79
    performance comparison, 79-83
  band, 8

V

Vee'd elements, 37
Vertical stacking, 183-185
Vhf
  antennas, 29-60
    commercial multiband, 41-60
      medium-signal, 44-50
      medium- to weak-signal, 50-53
      strong-signal, 42-44
      weak-signal, 53-58
    measured performance comparison, 58-60

Vhf—cont
antennas
multiband, 35-41
fm reception, 39-41
multimode directors, 39
multimoding techniques, 36-39
single-band antennas, 32-35
commercial, 34-35
log-periodic principle, 33-34
single-channel antennas, 29-32
high band, 8
low band, 8
-uhf antennas, 85-104
commercial, 88-104
measured performance comparison, 101-104
medium-signal, 94-100
strong- to medium-signal, 89-94
weak-signal, 100-101
isolation networks, 86-87
Volt, 10
Voltage systems, 10

W

Wall mounts, antenna, 156-157
Waveguide effect, 76
Wavelength and frequency, antenna, 17-18
Weak-signal
all-channel antennas, 100-101
vhf multiband antennas, 53-58
Wideband element, 23

Y

Yagi
corner-reflector, 69-74
couplers, 196-198